Current Chinese Economic Report Series

The Current Chinese Economic Reports series provides insights into the economic development of one of the largest and fastest growing economies in the world; though widely discussed internationally, many facets of its current development remain unknown to the English speaking world. All reports contain new data, which was previously unknown or unavailable outside of China. The series covers regional development, industry reports, as well as special topics like environmental or demographical issues.

More information about this series at http://www.springer.com/series/11028

Ling Wang • Shao-ju Lee • Ping Chen
Xiao-mei Jiang • Bing-lian Liu
Editors

Contemporary Logistics in China

New Horizon and New Blueprint

Springer

Editors
Ling Wang
Logistics Research Center
Nankai University
Tianjin, China

Shao-ju Lee
Logistics Research Center
Nankai University
Tianjin, China

Ping Chen
Business School
Nankai University
Tianjin, China

Xiao-mei Jiang
College of Management
Tianjin Normal University
Tianjin, China

Bing-lian Liu
Logistics Research Center
Nankai University
Tianjin, China

ISSN 2194-7937 ISSN 2194-7945 (electronic)
Current Chinese Economic Report Series
ISBN 978-981-10-1051-4 ISBN 978-981-10-1052-1 (eBook)
DOI 10.1007/978-981-10-1052-1

Library of Congress Control Number: 2016942095

Printed on acid-free paper

This Springer imprint is published by Springer Nature
The registered company is Springer Science+Business Media Singapore Pte Ltd.

Contents

Chapter 1
Development of China's Logistics Market

Xiao-mei Jiang

The domestic and overseas development environment of China's logistics industry was very complex in 2014. As the world's economy recovering slowly and China's domestic economy entering a new normal development stage, the overall growth rate of China's logistics market has slowed down to some extent. Meanwhile, the internet economy is being developed rapidly in China, the Chinese Government is vigorously promoting the "One Belt and One Road"[1] initiative, the Pilot Free Trade Zone and other major national development strategies; these actions have created new development potential for China's logistics industry.

In 2014, the major characteristics of China's logistics market are manifested in the following four aspects. First, the China-Europe freight rail lines, the cross-border e-commerce logistics, the urban distribution and rural logistics are developing rapidly, propelled jointly by the solid domestic demand and the favorable government policies. Second, China's logistics industry is highly regarded by the domestic and overseas investors; thus, online retail logistics, online food delivery, express and expedited delivery[2] and logistics property have become the investment hot spot. Third, the warehousing enterprises, express and distribution enterprises have actively brought in the mobile internet and big data technology to improve

[1] "One Belt and One Road" is the abbreviation of the "Silk Road Economic Belt and the 21st Century Maritime Silk Road;" it is an ideal and proposal for cooperative development pronounced by the Chinese leaders. Borrowing from the historical symbol of ancient Chinese "Silk Road," and basing on the bilateral and multilateral mechanism between China and relevant countries, the initiative aims to jointly build the community of interest, fate, and responsibility, with mutual political trust, economic integration and culture tolerance.

[2] Expedited delivery refers to the delivery mode in between the express and the ordinary deliveries by highway transport. The weight of transported cargo of this mode is generally 30 kg–300 kg. The transport speed is slower than that of the express delivery, but faster than that of the ordinary delivery.

X.-m. Jiang (✉)
College of Management, Tianjin Normal University, Tianjin, People's Republic of China
e-mail: macro04@126.com

L. Wang et al. (eds.), *Contemporary Logistics in China*, Current Chinese Economic Report Series, DOI 10.1007/978-981-10-1052-1_1

their operations efficiency. Fourth, the transformation and upgrading of China's logistics industry have been accelerated; this is reflected by (a) large highway logistics parks beginning to develop towards chain operations, (b) special-line enterprise[3] of less-than-truck-load building various transportation enterprise alliances, (c) large local express enterprises aggressively transforming to integrated logistics enterprises, and (d) China Railway Corporation vigorously developing new freight products.

This chapter is divided into three sections. Section 1.1 describes the economic and policy environments of China's logistics industry in 2014. Section 1.2 introduces the general status of China's logistics market, including specifically the total expense of social logistics, the added value of logistics industry, the freight volume and freight turnover, the cargo throughput and container throughput of port, and other indicators. Section 1.3 summarizes the major characteristics of China's logistics market in 2014.

1.1 China's Logistics Development Environment

1.1.1 Sluggish Recovery of World Economy

In 2014, the global economy was still in the adjustment period after the global financial crisis and recovering slowly. In the developed economies, the upswing of U.S. economy was relatively apparent, and the growth rates in Euro zone and Japan had fluctuated at lower level. The economic growth of developing economies in South Africa, Brazil and others was decelerated. Table 1.1 shows the economic growth of major global economic blocks for 2012–2014.

In 2014, the global trade exhibited a state of low growth owing to the lethargic global demand. According to the United Nations' estimate, the volume of world trade in 2014 was expected to increase by 3.4 %, which was slightly higher than the previous year's growth of 3 %, but was way lower than the 7.0 % average level prior to the international financial crisis. The Baltic Dry Index (BDI) of the international shipping market is the direct index representing the prosperity of the international trade. In 2014, the BDI was generally below the balance point (2,000 points) of maritime transport, and falling erratically. It had slid down to 782 points on December 24 from the 2,113 points on January 2, dropping by 62.9 % over 1 year. Figure 1.1 shows the BDI for 2014.

Affected by the slow recovery of global economy, the weakening of low cost advantage in China, the declined foreign capital investment in manufacturing industry, and the continual falling of bulk commodity prices in the international market,

[3] Special-line enterprises of less-than-truck-load mainly engage in transporting goods following pre-set lines between designated cities, and the transport batch size of the cargo is generally 300 kg–3000 kg.

Table 1.1 Economic growth of major global economic blocks for 2012–2014

			Unit: %
Country	2012	2013	2014
Global Average	2.4	2.5	2.6
USA	2.3	2.2	2.4
Eurozone	−0.7	−0.5	0.8
Japan	1.5	1.5	0.2
South Africa	2.5	1.9	1.4
Brazil	1.0	2.3	0.1
India	5.1	6.9	7.4
Russia	3.4	1.3	0.7
Mexico	4.0	1.4	2.1

Source: *World Economic Status: Retrospect for 2014 and Outlook for 2015*, the National Bureau of Statistics, the P.R.C., 2015-02-27
http://www.stats.gov.cn/tjsj/zxfb/201502/t20150227_686531.html

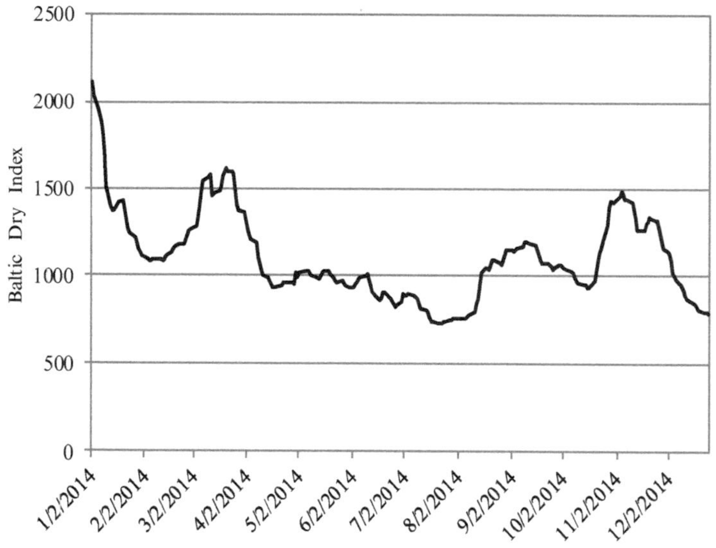

Fig. 1.1 Baltic dry index for 2014
Source: Finance.ifeng.com. Baltic Dry Index. http://app.finance.ifeng.com/data/indu/jgzs.php?symbol=58?; 2015-04-30

the overall growth rate of the China's foreign trade in 2014 had dropped markedly. The total annual import and export volume was 4.3 trillion USD, showing a year-on-year growth of 3.4 %; this growth rate receded by 4.2 percentage points from that of 2013. Figure 1.2 shows China's total amount of import and export and growth rate for 2006–2014.

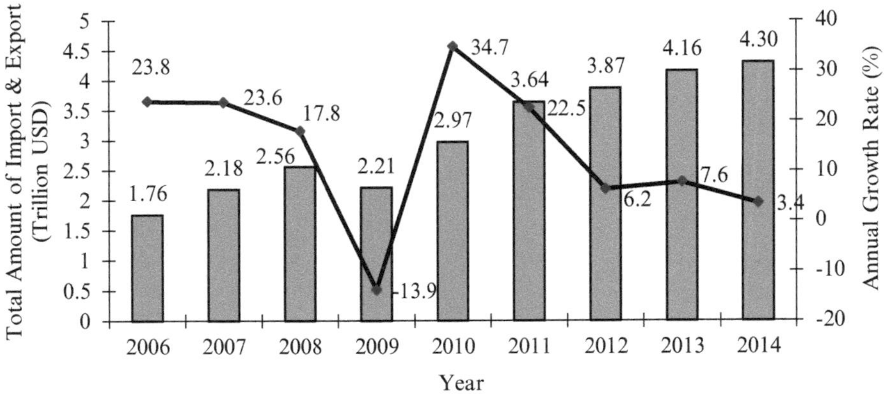

Fig. 1.2 China's total amount of import and export and growth rate for 2006–2014
Source: Compiled from the *China Statistical Yearbook (2014)*, and the *China Statistical Bulletin of National Economic and Social Development (2014)*, both published by the National Bureau of Statistics of China

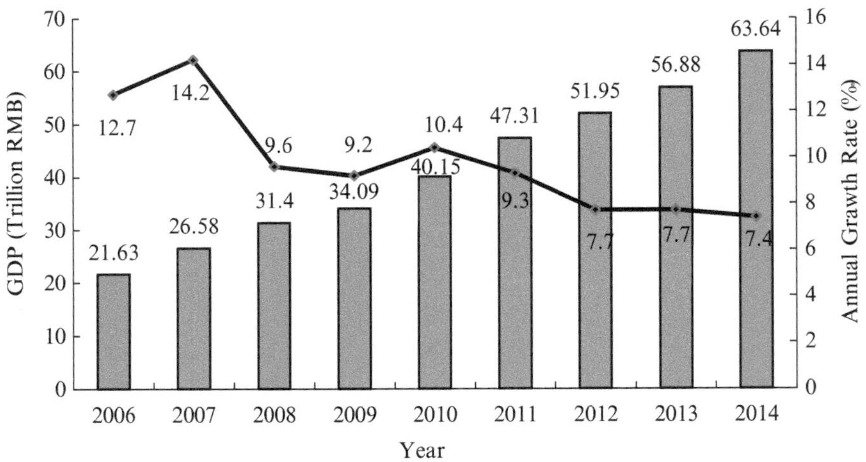

Fig. 1.3 China's GDP and growth rate for 2006–2014
Note: GDPs are computed per current price of each year; growth rates are computed based on constant price of the same base year
Source: Compiled from the *China Statistical Yearbook (2014)*, and the *China Statistical Bulletin of National Economic and Social Development (2014)*, both published by the National Bureau of Statistics of China

1.1.2 The "New Normal Stage" of China's Economy

In 2014, China's GDP (gross domestic product) was 63.64 trillion RMB, which represented a year-on-year growth of 7.4 %, computed in accordance with the constant price. Figure 1.3 shows China's GDP and growth rate for 2006–2014.

Table 1.2 Nine features of China's economy in the "new normal stage"

	Features
Consumer demand	The stage of waved emulative spending has basically ended; personalized and diversified consumption is gradually becoming the mainstream.
Investment demand	The traditional industries are relatively saturated; profuse investment opportunities for interconnected infrastructures, various new technologies, new products, new types of business, and new business modes are emerging.
Export and international balance of payment	"Bringing in" of high-level enterprises and "going abroad" on a large scale are occurring simultaneously.
Production capacity and mode of industrial organization	Capacity of traditional industries far exceeds the demand, so the industrial structure must be optimized and upgraded; merger and reorganization of enterprises and relatively more concentrated production are inevitable; the effect of emerging industry, service industry and small and micro-enterprises is more prominent; the miniaturization, intelligence and professionalization in production will become new features of the industrial organization.
Production factors	The advantage of low-cost labor force has weakened; economic growth will rely more on the quality of the human capital and technological advancement, therefore, innovation must become the new engine driving the development.
Market competition	Volume expansion and price competition are being gradually transformed into quality-oriented and differentiation-focused competition.
Resource and environmental constraints	The environmental bearing capacity has reached or is nearing the upper limit, therefore promoting the new form of green and low-carbon cyclic development is crucial.
Economic risk	It is controllable on the whole, but it will take some time to alleviate the various risks featured by high leverage and bursting bubble.
Resource allocation model and macro-control mode	The marginal effect of the comprehensive stimulation policy is obviously declining, therefore, the future industrial development direction must be explored through exercising the market mechanism.

Source: xinhuanet.com. The Central Economic Working Conference of the Communist Party of China was held in Beijing
http://news.xinhuanet.com/fortune/2014-12/11/c_1113611795.htm.2014-12-11

Seeing from the development features, China's economy had parted from the high-speed growth (above 9%) since 2012 and has entered the "New Normal Stage;" it has taken on many new features reflected specifically in nine aspects, as shown in Table 1.2.

The change in tendency after the economic development entered the New Normal Stage shows that, China's economy is evolving towards the stage of more advanced form, more complicated division of labor, and more reasonable structure. China's logistics industry needs to adapt to the altered characteristics of the new economic norm, so as to accelerate its development, and provide the support for promoting the transformation and upgrading of the economy and the service of the citizens' livelihood.

1.1.3 *Rapid Development of Internet Economy*

2014 marks the 20th year of China's full accesses to the internet. In recent 20 years, internet has developed and grown rampantly in China, and permeated continuously into all industries, thus, begetting a host of new products, new types of business and new business modes, and has profoundly changed the production and life style.

1.1.3.1 Mobile Application of Innovative O2O[4] and Abundant Market Opportunity

In 2014, China's mobile e-commerce experienced a robust development trend. Mobile internet, mobile payment, two-dimension code and other mobile internet technologies made marked breakthroughs in succession. The coupling of mobile internet technology with e-commerce has prompted the ushering of traditional industries in retail, catering, finance, tourism, medical treatment, education into the realm of internet; myriads of mobile applications and market opportunities of innovative O2O are produced.

The mobile shopping and catering O2O markets have grown rapidly. In 2014, China's online retail sales[5] reached 2.79 trillion RMB, which ranked the highest in the world for two consecutive years. This amount represents a year-on-year growth of 49.7 %, which is far higher than the 12.0 % growth rate of the total amount of retail sales of consumer goods. The growth in market size of mobile shopping was more than 200 %. The proportion of mobile terminal shopping accounts for 33.0 % of all online shopping, increasing by 19 percentage points from that of 2013 (iResearch Inc 2015). The market size of O2O in the catering industry reached 94.37 billion RMB, increasing by 51.5 % from that of 2013 (pintu360.com 2015). Figure 1.4 shows the scale and growth of China's mobile shopping market for 2011–2018.

1.1.3.2 Rapid Development of Cross-Border E-Commerce Market

After the international financial crisis, the demand of China's foreign trade had an apparent change. Export orders were changed from large orders and clumpy ordering to small orders and frequent ordering; domestic consumers showed increasing demands on high-quality imported products. Meanwhile, due to the popularization of internet and mobile internet, the disparity in commodity price at home and abroad and the quality information was diminished, giving rise to the basic conditions for cross-border transaction of small-sized and fragmented orders. These changes

[4] O2O, namely, Online To Offline, refers to the coupling of offline business opportunities and internet, so that the internet becomes the front desk of offline transaction.

[5] The online retail sales refers to the retail sales of the commodity and service realized through the transaction platform (including self-built websites and third-party platforms) on public network.

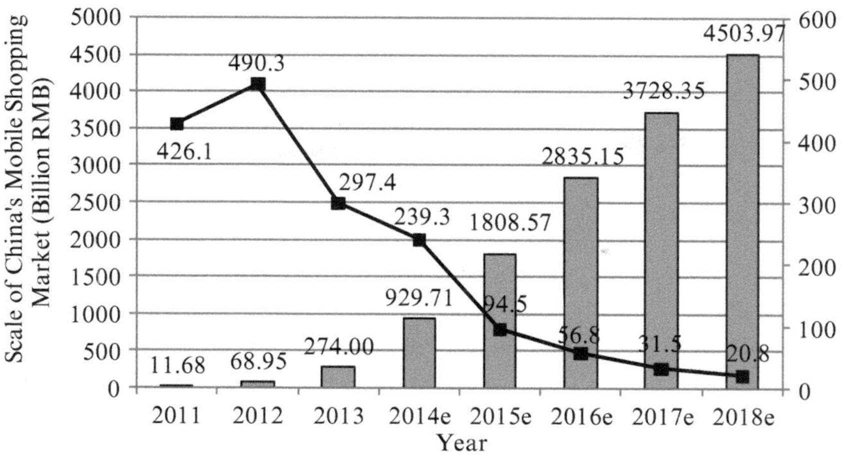

Fig. 1.4 Scale and growth of China's mobile shopping market for 2011–2018
Note: Salesfor 2014–2018 are estimated values
Source: iResearch Inc. "The scale of China's mobile shopping market is 929.71 billion RMB." http://report.iresearch.cn/html/20150202/245969.shtml. 2 Feb. 2015

together with improvement on cross-border payment, logistics, and Customs' supervision had jointly promoted the rapid development of the cross-border e-commerce market in China.

In 2014, the transaction scale of cross-border e-commerce in China was 4.2 trillion RMB, showing a year-on-year growth of 33.3 %. In which, cross-border export accounted for 85.4 %, and cross-border import accounted for 14.6 %; cross-border B2B trade accounted for 93.5 %, and B2C trade accounted for 6.5 % (China e-Business Research Center 2015). Cross-border e-commerce accounted for more than 10 % in the total volume of import and export trade, and became the prominent highlights of China's foreign trade.

1.1.3.3 Internet Promoted the Transformation of China's Manufacturing Industry

Internet greatly cuts down the information asymmetry between production and marketing, accelerates the close connection between the production end and the market end, and gives rise to the consumer-driven business mode. The consumer-driven mode requires that the production and manufacturing system to be highly flexible and personalized, and can respond swiftly to the market; these features are completely different from the standardized, batched, rigid and slow production mode under the traditional B2C business mode.

China's manufacturing industry is being transformed under the impetus of the internet, and under two typical paths. Under the first path, internet technology is used by the manufacturing enterprise to develop multi-variety, small-lot production and mass-customization. Take a furniture manufacturing company in Guangzhou as a typical case. With internet technology, the company built a C2Bmass-customization production mode similar to the "Dell Mode." First, this company provides the virtual design of three-dimensional personalized furniture for hundreds of offline stores through its powerful 3D design software and cloud design resource integration capability, so that customers can purchase the "Virtual Product" by ordering in the stores. Then, the company tallies the data of the "Virtual Products" ordered by the customers, and sends them to the manufacturing plant. Finally, the digitized manufacturing plant dissects and combines the lists of orders by parts, so as to realize scale production.

Under the second path, some supply chain service companies carryout the online connection and integration of the consumers, retailers and manufacturers with internet technology, so as to gain commonality among the supply chain subjects and realize flexible production on the supply chain. Take one supply chain management company in Guangdong as atypical case. This company, in response to the "multi-variety, small-lot and fast-response" orders from an e-commerce retail clothing enterprise, utilizes the digitized total-process, all-through, data-sharing supply chain, to integrate the e-commerce apparel supply chain, and realizes the bona fide flexible production of clothing. One typical application context is as follows. When an e-commerce retail clothing enterprise launches a new style, a batch of 50–100 items would be used to test the market. The e-commerce retail enterprise carries out the sales forecast on this style of clothing by analyzing the customers' data on clicking, collection, and shopping cart, and transmits the forecast data to the production factory in real time. Since the e-commerce retail enterprise is connected with the factory by the mutually built "Dynamic Replenishment-ERP" system, materials and capacity will be prepared in the factory according to the sales forecast and inventory status of the e-commerce retail enterprise. When the best-selling style is evidenced, the factory can provide the frequent and small-batched replenishments.

1.1.4 The "One Belt and One Road" Initiative and the Free Trade Zone Strategy

1.1.4.1 The "One Belt and One Road" Initiative Advocated by the Government

The "One Belt and One Road" initiative is a major initiative proposed by Chinese President Xi Jinping when he visited Central Asia and Southeast Asia countries in September and October of 2013, which was highly regarded by the international community. In 2014, China began actively promoting the "One Belt and One Road" Initiative, and had so far achieved notable progress.

China's top leaders have actively promoted the "One Belt and One Road" Initiative. President Xi Jinping, Prime Minister Li Keqiang and other state leaders had successively visited more than 20 countries, attended the dialogue meetings for strengthening the interconnection partnership, and the Sixth Ministerial Meeting of the China-Africa Cooperation Forum. They met with the heads of State and government officials of relevant countries for many times during their visits, expounded on the profound connotation and positive significance of the "One Belt and One Road" proposal with respect to the bilateral relationship and regional development, and had reached extensive consensus on the intention of co-building the "One Belt and One Road."

China took the lead to establish the Asian Infrastructure Investment Bank and the Silk Road Fund. In October 2014, China proposed to establish the Asian Infrastructure Investment Bank (AIIB). AIIB is positioned as a multilateral development mechanism in the Asian region among the governments, to support key infrastructure constructions and promote regional cooperation and partnership. Up to March 31, 2015, 43 countries in Asia, Europe, America, Oceania and Africa officially filed applications to join AIIB. In November 2014, China announced it would invest 40 billion USD to establish the "Silk Road Fund," as the strategic platform for China to implement capital export to the other countries, and to provide the investment and financing for infrastructure construction, resource development, and industry cooperation to countries along the "One Belt and One Road."

In March 2015, the National Development and Reform Commission (NDRC), the Ministry of Foreign Affairs and the Ministry of Commerce jointly issued the *Vision and Actions on Jointly Building the Silk Road Economic Belt and 21st Century Maritime Silk Road*, which stipulated the principle of co-construction, the framework and ideal, the highlights and mechanism for cooperation pertaining to the "One Belt and One Road" initiative.

In addition, China also signed the memorandum of cooperation of co-building the "One Belt and One Road" with some of the countries, and signed the memorandum on regional cooperation and border cooperation, as well as mid-term and long-term development plans with some adjacent countries. It has studied and compiled outlines of the regional cooperation plan with some adjacent countries, strengthened the communication and negotiation with the relevant countries along the Belt, and promoted some important cooperation projects on matters such as interconnection of the infrastructure, industrial investment, resource development, economic and trade cooperation, financial cooperation, cultural and educational exchange, ecological protection and maritime cooperation.

1.1.4.2 Further Promotion of the Free Trade Zone Strategy

After the China (Shanghai) Pilot Free Trade Zone was established in August 2013, the State Council approved expanding the area of the Shanghai Pilot Free Trade Zone from 28.78 km^2 to 120.72 km^2 in December 2014. To duplicate and popularize the experience of the Shanghai Pilot Free Trade Zone across the nation, it approved the establishment of three new FTZs in Guangdong, Tianjin and Fujian. It mandated that based on the contents of the Shanghai Pilot Free Trade Zone, the three new FTZs are to enrich the contents by each considering its own regional characteristics.

So far, the layout of the FTZs in the eastern coastal region corresponding to Pearl River Delta, Yangtze River Delta, Bohai Economic Rim and the Western Side of the Taiwan Straits in China has been completed. The four FTZs will take the lead in exploring the potential of the reform and solving the reform puzzles through building the new open economy system, leveraging the new regional economic cooperation mode, and constructing the legal business environment. In addition, based on the *Vision and Actions on Jointly Building the Silk Road Economic Belt and 21st Century Maritime Silk Road*, the cities that the free trade zones are located, i.e. Shanghai, Tianjin, Guangzhou, Shenzhen, Fuzhou and Xiamen, will also become part of the "One Belt and One Road," and especially as pacesetter and main force on the 21st Century Maritime Silk Road.

1.1.5 Two Regional Development Strategies

1.1.5.1 Advancement in the Beijing-Tianjin-Hebei Coordinated Development

The Beijing-Tianjin-Hebei region (Beijing and Tianjin Municipalities and Hebei Province) is one of the regions having the most development strength and potential in Eastern China. In February 2014, the State Council proposed the national strategy of the Beijing-Tianjin-Hebei Coordinated Development. By the end of 2014, the strategy has obtained many implementation achievements in the fields of traffic, industry and environment.

The regional transport integration was promoted in the early stage. The Ministry of Transport designated the leading group for promoting the transport integration in Beijing-Tianjin-Hebei region, and formulated the scheme for division of tasks. The governments of these three places also enhanced the construction of the regional intercity railways, perfected the regional expressway network, promoted the coordinated development of the ports and airports in the region, and advanced the interconnection of transportation infrastructures in the region.

The three places achieved positive progress in industrial transfer. Beijing has transferred more than 500 non-core investment projects[6] to Tianjin. Hebei Province brought in the industries transferred out from Beijing and Tianjin, and arranged 40 industrial parks to receive these industries.

The regional environmental protection was promoted orderly. The three places founded the alliance for energy conservation, environmental protection and low-carbon consumption, built the joint control mechanism for water pollution emergency, and established the vehicle emission pollution monitoring platform in Beijing-Tianjin-Hebei region and the surrounding areas.

[6] The development of Beijing Municipality is positioned as the nation's political center, cultural center, international exchange center, and technological innovation center. Any industry not conforming to the positioning of these four major centers is deemed as "non-core" industry, and will be successively transferred to Tianjin or Hebei.

1.1.5.2 Steady Promotion of the Yangtze River Economic Belt Development Strategy

The Yangtze River Economic Belt consists of eleven Municipalities and Provinces of Shanghai, Jiangsu, Zhejiang, Anhui, Jiangxi, Hubei, Hunan, Chongqing, Sichuan, Yunan, and Guizhou. The population and GDP of this Belt are more than 40 % of the nation's total. In March 2014, the State Council launched the national development strategy of Yangtze River Economic Belt.

Yangtze River channel is the most important east–west axis of the national territorial development in China, and holds an important strategic position in the overall pattern of regional development. Currently, China's economic development has entered the critical period of gradient migration from East to West. Building the Yangtze River Economic Belt, basing on the Yangtze River golden waterway, not only could directly drive the new development power of more than 1/5 of the nation's land and 600 million population, to close the discrepancies among the Eastern, Central and Western regions, but also could optimize the economic structure, and build the new economic supporting belt and new open cooperation platform with global influence.

In September 2014, the State Council also launched the guideline on developing the Yangtze River Economic Belt and transport planning. In the guideline, the Belt was positioned as: the inland river economic belt having global influence, the coordinated development belt of interactive cooperation among the Eastern, Central and Western regions, the comprehensively promoted internal and external open zone along the coast, the river and the bordering areas, and the advanced demonstration belt of ecological civilization establishment. It also advocated that the development of the Belt is promoted by improving the functions of the Yangtze River golden waterways, building the comprehensive three-dimensional traffic corridor, propelling the industry transformation and upgrading by innovation, comprehensively advancing the new-type urbanization, and cultivating the new advantage from complete opening to the external world. For traffic planning, it delineated the construction goals and key tasks of the comprehensive three-dimensional traffic corridor of the Belt.

1.2 The Overall Scale of China's Logistics Market

1.2.1 Total Value of Social Logistics

In 2014, the value of the total national social logistics[7] remained in the state of tepid growth. The total value of social logistics was 213.5 trillion RMB, which represents a growth of 7.9 % computed at constant price; this growth rate declined by 1.6 per-

[7]The value of the total national social logistics is composed of five parts: ① total value of commodities of farming, forestry, animal husbandry and fishing products in social logistics field ("total

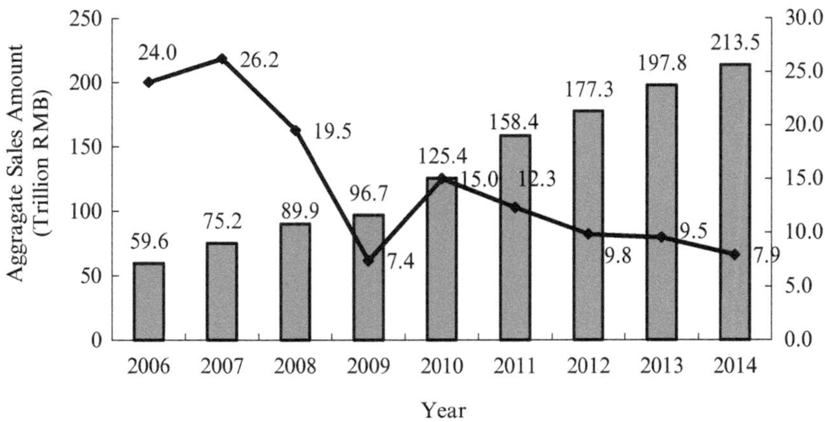

Fig. 1.5 Total value and growth rate of social logistics for 2006–2014
Note: the growth rates for 2006–2009 are nominal growth rates, and that for 2010–2014 are real growth rates
Source: Compiled from the *National Logistics Operations Bulletin for 2006–2014*, published by the National Development and Reform Commission, the National Bureau of Statistics, and the China Federation of Logistics and Purchasing

centage points from that of the previous year. Figure 1.5 shows the total value and growth rate of social logistics for 2006–2014. Table 1.3 shows the composition of total value of social logistics for 2009–2014.

1.2.2 Total Expense of Social Logistics

In 2014, the total national social logistics expense was10.6 trillion RMB, increasing by 6.9 % from that of last year. The total social logistics expense accounted for 16.6 % of the DGP, which was 0.3percentage points lower than that of previous year. The change of the share of total social logistics expense in GDP was affected by the freight volume, freight turnover, adjustment of the GDP data, and was also the result of structural change in China's economy. Table 1.4 shows the total logistics expense, the increment and the share in GDP for 2006–2014. Table 1.5 shows the composition of total expense of social logistics for 2014.

---

logistics amount of agricultural products" for short); ② total value of commodities of industrial products in social logistics field ("total logistics amount of industrial products" for short); ③ total logistics value of import goods; ④ total value of commodities of renewable resources in social logistics field ("total logistics amount of renewable resources" for short); ⑤ logistics value of goods from institutes and residents (including luggage in railway and air transport, parcels and letters in mailing service, various donations from all sectors of the society and handling, carrying and transportation of goods incurred by moving activities of institutes and residents.)

Table 1.3 Composition of total value of social logistics for 2009–2014

	2009		2010		2011		2012		2013		2014	
Indicator	Value (trillion RMB)	Share (%)	Value (trillion RMB)	Share (%)	Value (trillion RMB)	Share (%)	Value (trillion RMB)	Share (%)	Value (trillion RMB)	Share (%)	Value (trillion RMB)	Share (%)
Industrial products logistics	87.4	90.4	113.1	90.2	143.6	90.2	162.0	91.4	181.5	91.8	196.9	92.2
Imported goods logistics	6.9	7.1	9.4	7.5	11.2	7.1	11.5	6.5	12.1	6.1	12.0	5.6
Agricultural products logistics, etc.	2.4	2.5	2.9	2.3	3.6	2.7	3.8	2.1	4.2	2.1	4.6	2.2
Total	96.7	100.0	125.4	100.0	158.4	100.0	177.3	100.0	197.8	100.0	213.5	100.0

Source: Compiled from *the National Logistics Operations Bulletin for 2006–2014*, published by the National Development and Reform Commission, the National Bureau of Statistics, and the China Federation of Logistics and Purchasing

Table 1.4 Total logistics expense, increment and share in GDP for 2006–2014

Year	Total expense for social logistics cost (trillion RMB)	Increment from previous year (%)	Share in GDP (%)
2006	3.8	13.5	18.3
2007	4.5	18.2	18.4
2008	5.5	16.2	18.1
2009	6.1	7.2	18.1
2010	7.1	16.7	17.8
2011	8.4	18.5	17.8
2012	9.4	11.4	18.0
2013	10.2	9.3	16.9
2014	10.6	6.9	16.6

Note: In 2013, the Ministry of Transport adjusted the statistical scope of the freight volume and freight turnover, and the National Bureau of Statistics readjusted the total social logistics costs in 2013 and its share in GDP according to the freight volume and freight turnover under the new statistical scope. Values in this table are adjusted data

Source: Compiled from the *National Logistics Operations Bulletin for 2006–2014*, published by the National Development and Reform Commission, the National Bureau of Statistics, and the China Federation of Logistics and Purchasing

Table 1.5 Composition of total expense of social logistics in 2014

Indicator	Value (trillion RMB)	Increment from previous year (%)	Share (%)
Total expense of social logistics	10.6	6.9	100.0
Including:			
Transportation expense	5.6	6.6	52.9
Storage expense	3.7	7.0	34.9
Management expense	1.3	7.9	12.2

Source: Compiled from the *National Logistics Operations Bulletin for 2014*, published by the National Development and Reform Commission, the National Bureau of Statistics, and the China Federation of Logistics and Purchasing

1.2.3 Added Value of Logistics Industry

In 2014, the nation's total added value of logistics industry was 3.5 trillion RMB, with a year-on-year growth of 9.5 %. Added value of the transportation industry was 2.4 trillion RMB, showing a year-on-year growth of 8.3 %; added value of the trade logistics industry was 678.1 billion RMB, representing a year-on-year growth of 7.9 %; added value of warehousing industry and postal logistics industry increased by 4.8 % and 35.6 %, respectively. Table 1.6 shows the nationwide logistics added value and growth rate for 2006–2014.

Table 1.6 Nationwide logistics added value and growth rate for 2006–2014

Year	Added value of logistics industry (trillion RMB)	Growth rate (%)
2006	1.4	12.5
2007	1.7	20.3
2008	2.0	15.4
2009	2.3	7.3
2010	2.7	13.1
2011	3.2	13.9
2012	3.5	9.1
2013	3.9	8.5
2014	3.5	9.5

Note: In 2013, the Ministry of Transport adjusted the statistical scope of the freight volume and freight turnover, and the National Bureau of Statistics readjusted the total social logistics costs in 2013 and its share in GDP according to the freight volume and freight turnover under the new statistical scope. Values in this table are adjusted data

Source: Compiled from the *National Logistics Operations Bulletin for 2006–2014,* published by the National Development and Reform Commission, the National Bureau of Statistics, and the China Federation of Logistics and Purchasing

1.2.4 Freight Volume and Freight Turnover

In 2014, the freight volume completed by China was 43.91 billion tons, an increase of 7.1 % comparing with that of last year. The freight turnover was 18.46 trillion tons-km, which was increased by 9.9 % from the previous year. Table 1.7 shows the annual nationwide freight volume, freight turnover, and growth rate for 2006–2014.

In 2014, the railway transport was affected by the decelerated growth of relevant industries and diminishing cargos for bulk goods, the freight volume and freight turnover declined substantially; several other transportation modes had faster growth. Table 1.8 shows the freight volume, freight turnover and growth rate of various transportation modes in 2014.

1.2.5 Cargo Throughput and Container Throughput of Port

In 2014, affected by the slower growth of the macro economy and the adjustment of economic structure, the growth rates of cargo throughput and container throughput of port showed evident drop from previous year. Cargo throughput completed for the nationwide ports was 12.45 billion tons, showing a year-on-year growth of 5.8 %; this growth rate receded by 3.4 percentage points. The foreign-trade cargo throughput was 3.59 billion tons, representing a year-on-year growth of 6.9 %; this growth rate fell by 3.0 percentage points. Table 1.9 shows the nationwide cargo throughput and growth rate of port for 2006–2014.

Table 1.7 Annual nationwide freight volume, freight turnover, and growth rate for 2006–2014

Year	Freight volume		Freight turnover	
	Volume (billion ton)	Growth rate (%)	Freight turnover (trillion ton-km)	Growth rate (%)
2006	20.37	9.40	8.88	10.70
2007	22.76	11.70	10.14	14.20
2008	25.86	13.60	11.03	8.80
2009	28.25	9.20	12.21	10.70
2010	32.41	14.70	14.18	16.10
2011	36.97	14.10	15.93	12.30
2012	41.00	10.90	17.38	9.10
2013	40.99	—	16.80	—
2014	43.91	7.10	18.46	9.90

Note: In 2013, the Ministry of Transport conducted the special investigation of the transportation industry, and adjusted the statistical scope of highway and waterway transport volume. In the table, data before 2013are based on the original statistical scope, and data for2013 and 2014 are based on the new statistical scope

Source: Compiled from the *China Statistical Yearbook (2014)*, and the *China Statistical Bulletin of National Economic and Social Development (2014)*, both published by the National Bureau of Statistics of China

Table 1.8 Freight volume, freight turnover and growth rate of different transportation modes in 2014

Indicator	Volume	Growth rate (%)	Share (%)	Change of share (% point)
Total freight volume (billion ton)	43.91	9.90	—	—
Among which:				
Railway	3.81	−3.90	8.70	−0.10
Highway	33.43	8.70	76.10	−2.70
Waterway	5.96	6.40	13.60	2.70
Civil aviation (megaton)	5.93	5.70	—	—
Pipeline	0.69	5.20	1.60	0.10
Freight turnover (trillion ton-km)	18.46	9.90	—	—
Among which:				
Railway	2.75	−5.60	14.90	−2.50
Highway	6.11	9.70	33.10	−0.10
Waterway	9.19	15.70	49.80	2.50
Civil aviation (billion ton-km)	18,610	9.30	—	—
Pipeline	0.39	10.90	2.10	0.00

Source: Compiled from the *China Statistical Yearbook (2014)*, and the *China Statistical Bulletin of National Economic and Social Development (2014)*, both published by the National Bureau of Statistics of China

Table 1.9 Nationwide cargo throughput and growth rate of port for 2006–2014

Year	Nationwide cargo throughput of port		Share of cargo throughput of foreign trade	
	Volume (billion ton)	Growth rate (%)	Volume (billion ton)	Growth rate (%)
2006	5.57	14.80	1.61	18.10
2007	6.41	15.10	1.85	14.60
2008	7.02	9.60	1.99	7.40
2009	7.66	9.00	2.18	9.80
2010	8.93	16.70	2.50	14.70
2011	10.04	12.40	2.79	11.40
2012	10.78	7.30	3.06	9.70
2013	11.77	9.20	3.36	9.90
2014	12.45	5.80	3.59	6.90

Source: Compiled from the *Statistical Bulletin of Road and Waterway Transportation Industry for 2006–2014*, published by the Ministry of Transport

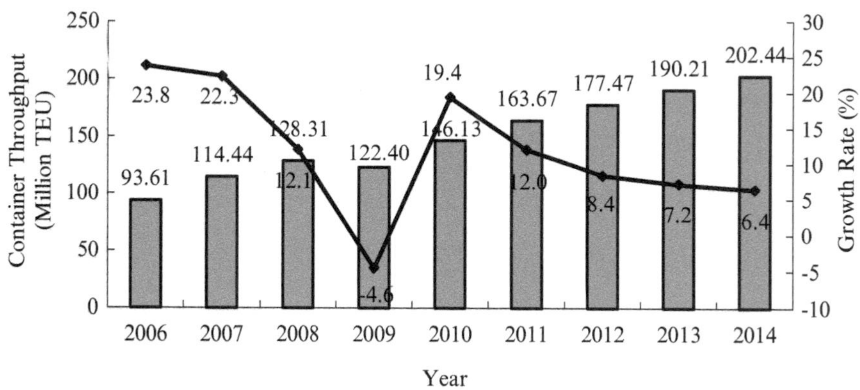

Fig. 1.6 Nationwide container throughput and growth rate for 2006–2014
Source: Compiled from the *Statistical Bulletin of Road and Waterway Transportation Industry for 2006–2014*, published by the Ministry of Transport

In 2014, the container transport market kept the steady growth momentum. The annual container throughput completed in the nationwide ports was 202.44 million TEUs, which increased by 6.4 % comparing with that of last year; this growth rate dipped 0.8 percentage points from that of previous year. Figure 1.6 shows the nationwide container throughput and growth rate for 2006–2014.

1.2.6 Cargo and Mail Throughput for Civil Airport

In 2014, the cargo and mail transport of the nation's airports continued the trend of steady growth, with a higher growth rate. The total annual cargo and mail throughput completed was 13.561 million tons, which was increased by 7.8 % comparing

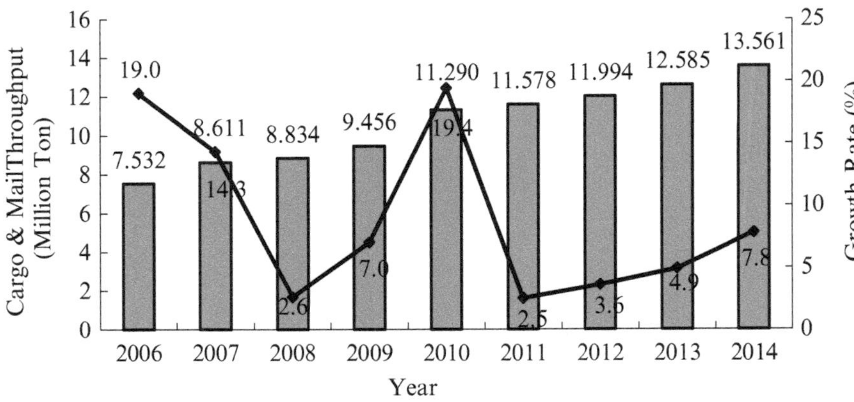

Fig. 1.7 Nationwide cargo and mail throughput and growth rate in civil airports for 2006–2014
Source: Compiled from the *Statistical Bulletin on the Productivity of National Airports for 2006–2014*, published by the Civil Aviation Administration

with that of previous year; the growth rate was 2.9 percentage points higher than that of 2013. Figure 1.7 shows the nationwide cargo and mail throughput and growth rate in civil airports for 2006–2014.

1.2.7 Business Volume of Express Industry

In recent years, driven by the rapid development of e-commerce, the business volume of express industry in China has kept a rapid pace of growth. In 2014, the annual business volume completed by the above-scale express enterprises was 13.96 billion items, chalking a year-on-year growth of 51.9 %. Figure 1.8 shows the nationwide business volume of express delivery enterprises above the designated scale and the growth rate for 2008–2014.

The business structure and regional distribution of the express market in China have basically been stable in recent years. As for the business structure, the proportion of within-city business volume was kept within 25 %; the proportion of out-of-city business volume remained above 70 %, except for 2008; the proportion of business volume for international, Hong Kong, Macao and Taiwan dropped slightly. As for the regional distribution, the proportion of the business volume in the Eastern region maintained at about 80 %, the proportion of the business volume in the Central region remained at about 10 %, and the proportion of the business volume in the Western region had a slight decrease. Table 1.10 shows the business volume by categories of express delivery for 2008–2014. Table 1.11 shows the regional distribution of express delivery business for 2008–2014.

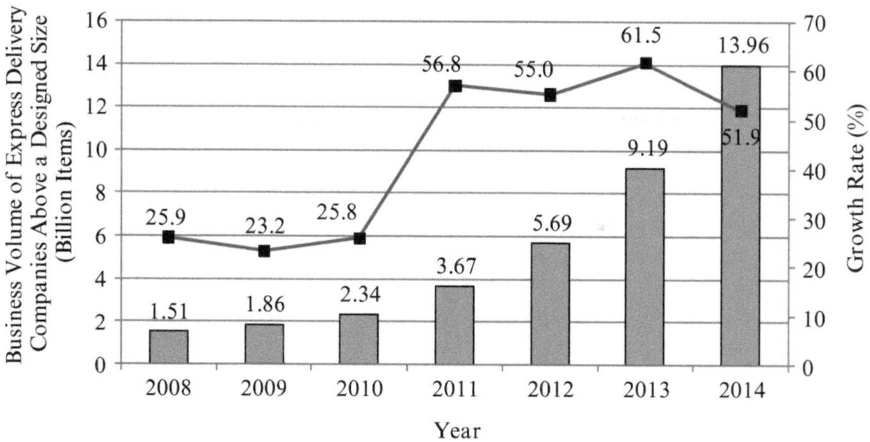

Fig. 1.8 Nationwide business volume of express delivery enterprises above the designated size and the growth rate for 2008–2014
Source: Compiled from the *Statistical Bulletin of Postal Service Industry for 2008–2014*, published by the State Post Bureau

Table 1.10 Distribution of business volume of express delivery by category for 2008–2014

			(Unit: %)
Year	Within-city	Out-of-city	International and Hong Kong, Macao and Taiwan
2008	26.6	66.4	7.0
2009	23.5	70.4	6.1
2010	22.9	71.5	5.6
2011	22.3	74.2	3.5
2012	23.1	73.7	3.2
2013	24.9	72.3	2.8
2014	25.4	72.3	2.3

Source: Compiled from the *Statistical Bulletin of Postal Service Industry for 2008–2014*, published by the State Post Bureau

Table 1.11 Distribution of business volume of express delivery by region for 2008–2014

Year	Eastern region	Central region	Western region
2008	78.5	11.5	10.0
2009	80.2	10.6	9.2
2010	79.3	11.6	9.1
2011	79.9	11.2	8.9
2012	81.9	10.5	7.6
2013	81.3	10.8	7.9
2014	82.0	10.6	7.4

Source: Compiled from the *Statistical Bulletin of Postal Service Industry for 2008–2014*, published by the State Post Bureau

1.3 Major Characteristics of China's Logistics Market

1.3.1 Development of Cross-Border Logistics System

1.3.1.1 China-Europe Freight Rail Lines

After the "One Belt and One Road" strategic concept was proposed, provinces along the paths in China actively embraced this strategy, and opened up a batch of China-Europe freight rail line[8] within a short period of time. Currently, China-Europe freight rail lines have had three channels; they are, respectively, the Eastern Channel departing from the Manzhouli port of Inner Mongolia autonomous region, the Central Channel departing from the Erenhot port of Inner Mongolia autonomous region, and the Western Channel departing from the Alashankou port of Xinjiang Autonomous Region. The originating points of China-Europe freight rail lines in China include Chengdu, Chongqing, Wuhan, Zhengzhou, Suzhou, and Harbin cities. Chinese commodities departing from these originating cities can arrive at Moscow, Warsaw, Duisburg, Madrid and other European cities in about 15 days; this transport time is about one-third of that of marine transport.

According to the statistics of China Railway Corporation, a total of 308 arrays of China-Europe freight rail cars, hauling 26,070 containers, were set forth in 2014, which was 228 arrays more than that of the same period of last year, and an increase by 285 %. But while the China-Europe freight rail lines were developing rapidly, problems such as duplicate lines and shortage of goods on return trips were present. Therefore, China Railways Corporation organized, and coordinated the unified management on the lines, for the purpose of attaining the unification of "Brand name," "Transportation organization," "Whole course pricing," "Service standards," "Management team," and "Coordination platform."

1.3.1.2 Cross-Border E-Ccommerce Logistics

In order to adapt to the rapid development of cross-border e-commerce in China and improve the service level of cross-border e-commerce logistics, many local express enterprises in China actively developed international special logistics lines in 2014, and opened the transshipment service of overseas shopping. For example, China Postal Express opened four special lines (China-Kazakhstan-Russia) for international small packages. STO Express opened the special line to Japan, and cooperated with Pony Express and TNT Post to enter Europe. SF Express opened the special lines for small packages to Russia, and to Europe. The European Express service center and American service center of YUNDA Express officially began operation. YTO Express started the special line of "Russian Easy Mail." Best

[8] China-Europe freight rail lines refer to the fast freight line from China to Europe, suitable for transporting freight containers.

Express opened its complete line of global express business. SF Express, China Post Express and YUNDA Express opened the US-China transshipment service, to provide consulting and inquiry services regarding warehousing, transshipment and cross-border logistics transportation in the US, for domestic online shopping customers in China and overseas Chinese.

The local large-scale cross-border e-commerce enterprises in China also accelerated the pace in setting up "overseas warehouses." For example, Light in the Box in Europe began its overseas warehouse operation at the beginning of 2014, and it then opened overseas warehouses in North America at the beginning of 2015. At the beginning of 2015, AliExpress, which is the online trading platform subordinated to the Alibaba Group, also added nine overseas delivery depots in the US, England, Spain, France, Germany, Italy, Russia and Indonesia; sellers on AliExpress platform can provide local logistics service for consumers in these countries and neighboring areas.

1.3.2 Progress in Urban Distribution and Rural Logistics System

1.3.2.1 Urban Distribution

Since China's Ministry of Commerce and Ministry of Finance started the pilot test of joint distribution in urban areas in 2013, the three-tier distribution network, composed of distribution centers of logistics parks, public distribution centers and end-node distribution sites, has been preliminarily built in the first batch of 22 pilot cities; a batch of urban joint distribution information service platform was built, and a batch of urban joint distribution modes were implemented. Typical urban joint distribution include the following modes: (1) integrated trunk line and branch line mode developed by relying on logistics parks; (2) merchant's joint distribution mode developed by relying on professional wholesale markets; (3) "Online ordering and store pick-up (delivery)" mode developed by relying on community convenience stores, property management companies, and universities; (4) joint distribution mode developed by utilizing self-service delivery container cabinets.

The innovation of "last-mile delivery" of urban distribution has emerged in endless forms, and the "last-mile delivery" platforms of dozens of different modes were successively completed, such as SF Heike, Caijiao station, sposter. Popular "last-mile delivery" platform of O2O includes shbj.com, beequick, meishisong.

1.3.2.2 Countryside Logistics

In 2014, e-commerce giants Alibaba and Jingdong accelerated their layout of the rural e-commerce market. Under the impetus of rapid development of rural e-commerce, the two-way logistics network which sends the agricultural products to the city and ships industrial products from the city to the countryside has been vigorously developed.

The Alibaba Group invested 2.21 billion RMB in Qingdao RRS Logistics in December 2013, so as to become the main logistics service provider of major household appliances on its subordinated Tmall website, and assist the Tmall merchants to expand their business to third-tier and fourth-tier cities which were difficult to penetrate previously. RRS has the sole large parcel logistics service network in China having the capability of delivering goods to the countryside plus the installation service for appliances. By the end of 2014, RRS has formed the complete logistics network composed of 90 logistics distribution centers, more than 2,800 county-level logistics distribution stations and more than 17,000 service provider posts. It has nearly 2 million m^2 of warehouse space and can achieve full coverage of 2,600 districts and counties within the country; it can realize "delivery within 24 h" in more than 155 districts and counties, and "deliver within 48 h" in 460 districts and counties.

In 2014, Jingdong rapidly expanded its county-level logistics distribution network by building the "Pioneer station" and the "Jingdongbang" service stores. "Pioneer station" is Jingdong's self-run website; its couriers are selected from its current cadre of employees of urban distribution in first-tier and second-tier cities. These distribution employees are professionally trained and can undertake both merchandise sales and delivery work. In 2014, Jingdong founded a total of 192 Pioneer stations, added the coverage of 600 new districts and counties, making the total coverage of districts and counties to 1,900, which accounted for half of the total number the districts and counties in the country. "Jingdongbang" service shop adopts the franchise mode, and is mainly responsible for the distribution and installation of Jingdong's major appliances in the countryside. The first "Jingdongbang" service shop was opened in November 2014, and it will reach 1,000 service shops by the end of 2017.

In addition, the local express delivery and expedite service enterprises in China are also accelerate their layout of rural network. In 2014, 5,000 new national rural express service nodes were added, which covered 48 % of villages and towns in China. The leading enterprise of the local Less-than-Truck-Load Express companies, i.e. Deppon Express, added about 1,000 direct sales nodes in 2014, which were mainly set up in the third-tier and fourth-tier cities.

1.3.3 High Awareness of Investors on China logistics

1.3.3.1 Huge Overseas Financing for Large-Scale E-Commerce Enterprises in China

In May 2014, JD.com Inc was listed in NASDAQ in the US, and capitalized at 1.78 billion USD. Logistics is one of the core competitiveness of Jingdong, and logistics system construction is one of the main objects of its IPO financing. In

2014, Jingdong substantially improved the coverage of its logistics network in the country's lower-tier cities and rural areas. Up to the end of 2014, Jingdong had owned seven logistics centers nationwide, operated 123 mega-warehouses in 40 cities across the country, and built 3,210 delivery stations and pick-up stations, covering 1,862 districts and counties across the country. The number of districts and counties covered by its "211 Program"[9] service reached 134, and that of its "Next Day Delivery" service was increased to 866. After the listing, the number of its mega-warehouses was increased by 43, the number of delivery stations was up by 1,376, and the districts and counties covered by "211 Program" and "Next Day Delivery" service were increased by 91 and 610.[10]

In September 2014, Alibaba Group Holding Ltd. was listed in New York stock exchange, capitalized at about 25.03 billion USD, which was the world's highest IPO value in history. Logistics is the foundation of e-commerce, and also the element that Alibaba needs to enhance. Alibaba has proposed the "China Smart Logistic Network" development strategy in 2013. The goal of the project was to build a set of open and public warehousing facility network on a national scale, with the Internet of Things, cloud computing, e-finance and other new technologies through self-building, co-building, and cooperation in about 8–10 years, so as to provide logistics service for various B2B, B2C and C2C enterprises. In 2014, Alibaba became the shareholder of KXTX and RRS, which was the specific act in advancing this project.

1.3.3.2 Support of Risk Investors on Construction of Online Catering Distribution System

In 2014, the online catering O2O mode in China received continual support from the risk investors and developed rapidly. Catering enterprises capable of providing logistics service for the consumers became the main focus of risk investment. The reason for this is that such catering enterprises can effectively control the food safety and service quality through its own distribution system, thus have better development prospect. Financing of some online catering enterprises and their offline distribution system construction in 2014 are shown in Table 1.12.

[9] Under the "211 Program," JD.com ensures that orders received by the morning deadline (11:00 AM) will be delivered on the same day, and orders received by the evening deadline (11:00 PM) will be delivered by 3:00 PM on the following day.

[10] Based on Jingdong's IPO prospectus, it had 86 mega-warehouses, 1,620 delivery stations and 214 pick-up stations in 36 cities in February 2014; it offered the "211 Program" in 43 cities, and the "Next Day Delivery" service in another 256 cities.

Table 1.12 Financing of some online catering enterprises and their offline distribution system construction in 2014

Online catering enterprise	Financing amount	Status of offline distribution system construction
Meishisong	Tens of millions USD	Offers delivery service with own professional team to homes in five cities, including Beijing and Shanghai.
DAOJIA.COM	50 million USD	Provides distribution service in eight cities including Beijing and Shanghai; employs about 1000 couriers.
SHBJ.com	50 million RMB	Provides distribution service in nine cities including Beijing and Shenzhen; employs about 7000 couriers.
dianwoba.com	10 million USD	Provides just-in-time delivery service for restaurant takeout and supermarket goods in Shanghai and Hangzhou.
etaoshi.com	20 million USD	Provides distribution service in five cities including Beijing and Shenzhen; employs about 200 couriers.

1.3.3.3 Investment Expansion for Operations Network of Express and Expedite Delivery Enterprises

In 2014, many express and expedite delivery enterprises received large sum of capital investment. An emerging express enterprise, Quanfeng Express, received investment of several hundred million RMB, to be used mainly for the network construction and operations layout. ZJS Express, having more than ten years of development history in express delivery, received a strategic investment of 1 billion RMB, to be used mainly on the expansion of its domestic distribution network and overseas express network.

One highway transportation network platform company KXTX received investment of several hundred million RMB from Alibaba as a shareholder, thus greatly accelerated the development of its operations network. In 2014, its hub centers were increased to 21 from 9 in 2013, the franchising sites were increased from 300 sites to 1,300 sites, and the intercity point-to-point direct shipment freight lines were increased from 1,000 to 9,000. Another highway transportation network platform company, ANE Logistics Ltd., received investment of 50 million USD, to be used mainly for distribution center expansion, network line expansion, human resource cultivation, and IT system upgrading. In 2014, the network layout of this company was markedly expanded; its distribution centers were increased from 20 sites in 2013 to 80 sites, and its websites were increased from 750 sites to 2,000 sites.

1.3.3.4 Development and Promotion of Logistics Apps Driven by Investment

Influenced by the great success of Uber in the taxi field, a large amount of capitals were stimulated to enter China's logistics field in 2014, which greatly promoted the development of Apps in the logistics field. The financing and service mode of some logistics Apps in 2014 are shown in Table 1.13.

Table 1.13 Financing and service mode of some logistics Apps in 2014

Logistics app	Financing amount	Service mode
Kuaishou	3 million RMB	The mobile internet express distribution service platform provides delivery of urgent mails for the users with LBS (location based service). This application can effectively deploy the deliverymen in the area that the addresser is located, and the deliverymen can conveniently select their closest orders.
loji.com	5 million RMB and 3 million USD	The logistic mobile information platform provides logistics inquiry, issuance of special logistics information and other services for the logistics companies, transport companies, distribution stations, car owners, and cargo owners.
OTMS	6 million USD	SaaS-based community-type transportation management platform provides an information platform that the cargo owners, transportation service providers at different levels and the drivers can share.
GoGoVan	16.5 million USD	Provides truck abutting services in P2P mode so that truck drivers are connected with the users using the on-demand delivery service; free for both the users and drivers. The service is provided in six regions, i.e. China Mainland, Hong Kong, Taiwan, Singapore, Australia and Korea.
sd-express	About 10 million RMB	The intelligent dispatching platform of intra-city freight transport provides the intelligent taxi service for the intra-city freight transport with mobile internet.
ishansong.com	Several million RMB	The platform of direct delivery of designated person and rapid intra-city direct delivery service provides a faster designated person delivery service. The entire process, from the point the flash deliveryman receives the goods till he delivers to the receiving party, is monitored. After the user places an order, the system will automatically send the task to the mobile phone of the flash deliveryman who is closest to the user, and the flash deliveryman will immediately visit the user to provide the service.
1HHD.com	Several million RMB	Free mobile application for matching freight trucks and users. It connects the real-time position information of the idle truck, and publishes the real-time freight transportation demand of the owner of goods, so that the truck makes direct contact with the owner of goods, thus, reducing the intermediate links, and resolving the problems in finding a truck and paying high price.
lanxiniu.com	33 million RMB	The distribution platform of the intra-city truck delivery, through franchising, integrates the vehicle and driver resource in the community to realize the intra-city distribution.
		The users can enter the starting and ending locations, vehicle type, number of people and other demands; and then, the platform will give out the corresponding price, so that the users can confirm and call the truck. It pushes the demand to the drivers near the users, so the drivers can accept the order by themselves. It has opened the services in seven cities including Beijing, and Shanghai.

1.3.3.5 Large Capital Investment in Logistics Real Estate

In February 2014, Global Logistic Properties Ltd. received about 15.2 billion RMB investment from a Chinese-funded consortium, including China Life Insurance Company Ltd., China Development Bank International Holdings Ltd., and HOPU Investments. First stage investment of 9.7 billion RMB has been fully appropriated in June 2014. By virtue of this financing, Global Logistic Properties Ltd. has maintained its leading position in China's logistics real estate development field.

In May 2014, E-shang Warehousing Logistics Co. received the equity investment of 650 million USD from Dutch pension fund APG Asset Management Co. which is the largest pension fund assets management firm in Holland. E-shang Warehousing Logistics Co. is a professional company which provides development and warehousing facilities service for the e-commerce and retail industry, and offers development, operations and management services of logistics, production and office facilities for enterprise customers. It has developed customers such as Amazon, JD, Newegg, Kerry Logistics, and H&M. E-shang and APG Asset Management will build a joint venture company to develop and manage the standardized and modern logistics real estate throughout China.

1.3.4 Promoting China's Logistics Operations Efficiency with Internet Technology

1.3.4.1 Logistics Service Level of E-Commerce with Cloud Warehouse

Cloud warehouse service refers to the online-offline integrated order fulfillment service provided to customers by the logistics enterprises which perform logistics data analysis, intelligent order sorting, and optimal delivery consolidation, based on big data, cloud computing and other information technologies.

Cloud warehouse service in China was first brought forth by Best Logistics Company. Best Logistics, by connecting the big data of Taobao.com, analyzes the historic operating data of its e-commerce retail enterprises, and by combining the product attribute of the e-commerce retail enterprises, helps it to build the nationwide parallel branch warehouse network. Best Logistics also devices optimal placement of cargos in the branch warehouse according to the optimal paths chosen by the artificial intelligence and big data analysis technology. On the eve before sales festivals, merchandises will be prepared in the branch warehouse in advance, so as to reduce the distribution cost of order fulfillment and improve the distribution efficiency. Best Logistics can also automatically select the most appropriate express company for sending the goods according to the analysis and calculation of mass data, to ensure timely delivery. Up to September 2014, Best Logistics has had nearly 70 branch warehouses in 30 central cities across the country, which occupied an operational area of more than 1 million m^2, and serviced more than 100 e-commerce retail enterprises at home and abroad. In June 2014, China Postal Express also released similar cloud warehouse product for its e-commerce customers.

1.3.4.2 Express Delivery Level Improved with Internet Technology

In 2014, well-known domestic express enterprises in China, such as SF Express, China Postal Express, STO Express, YTO Express and YUNDA Express, had all released the online inquiry and express mail delivery service based on Apps, Wechat and Web websites, in an attempt to reduce the cost of sending and internal dispatching management by way of internet. Some new business modes based on the internet are also applied in the express field. For example, GT Company initiated the O2O mode on express delivery by employing the cloud computing, big data and mobile internet technology, and combining the Apps, Web clients and offline express agent sites in the express field, to provide the users a good experience in inquiring, sending and managing their express parcels. RR Express tries to adopt the "Crowd-sourcing+O2O"mode to render the intra-city express service with idle personnel. Kuaishou, Shansong and others improve the service level of intra-city express delivery by linking the online and offline operations through the mobile phone Apps.

1.3.4.3 Integration and Utilization of Logistics Resource Promoted by Freight Apps

At the end of 2013, freight Apps began to sprout in China, and entered into the outbreak period in 2014; a large number of freight Apps came online in droves. According to incomplete statistics, more than 200 freight Apps had appeared on the market. The main function of these freight Apps is to build the information exchange bridge for the cargo owners and drivers with the mobile internet, to promote the matching between trucks and cargos, and improve the resource integration and utilization level of the freight field. Some main types of freight Apps are shown in Table 1.14.

1.3.5 Acceleration of Transformation and Upgrading of Logistics Industry

1.3.5.1 Accelerated Development of Large Logistics Park Chains

Zhejiang Chuanhua Logistics vigorously expanded its intelligent highway port network in many regions across the country in 2014. Intelligent highway port is the business mode of the third-generation logistics park launched by Chuanhua

Table 1.14 Major types of freight App

Type	Functions	Typical product
Apps in logistics parks	Both cargo owners and drivers can use it. Cargo owners can publish and manage the source of goods, and select the drivers. Drivers can publish the source of cars and find the goods. Generally, such Apps do not involve in the specific transaction, but only provides the information interaction function, and third-party payment, insurance, oiling and other value-added service.	Apps of Linan Logistics and Chuanhua logistics
Apps for cargo owners	Only applicable to cargo owners. Provide query, ordering, tracking, account checking, inquiry and other functions for the cargo owners.	Chewang platform, Kaopu platform, Lulu Dispatch, OTMS, Yunmanman, and Good logistics
Apps for truck management	Provided for logistics companies having own vehicles or needing to manage out-sourced vehicles. Have management functions of vehicle archiving, vehicle dispatching, tracking, etc.	G7 truck management version, Lulu truck management, and Aidijie motorcade
Apps for Drivers	Provided for the drivers. Functions include publishing the source of trucks, finding the goods, feedback of the transportation process, etc.	G7 freight personnel, Guanchebao, Haoduoche, Yunmanman, and Kuaile cart
Apps for transportation enterprises	Mainly provided for use by cargo owners or potential cargo owners. Functions include order placing and delivery, inquiry and tracking, account checking and settlement, business advertisement, etc.	Apps of Huayu and Deppon
Apps for in-city distribution	Provides intra-city order placing and delivery by standard vehicle type.	Blue Rhino, No.1 truck, No.1 Huode, Shendun Express and Huolala

Logistics. It integrates the highway freight transport resources and builds the platform by employing the Internet of Things technology, the Cloud Computing Network technology and the management principles of e-commerce. The platform will enable medium and small-sized logistics enterprise to improve the logistics operation level and efficiency, accelerate the pace of allocation cargos to empty returned vehicles, and help the manufacturing enterprises to significantly reduce the logistics cost. Up to the end of 2014, Chuanhua highway port project has decided to set up in Qingdao, Tianjin, Chongqing and some 20 cities, wherein some projects have started to build, and a few projects have been built and put into operation.

After many years of exploration and practice, the Guangzhou Linan Logistics Group has innovated the Logistics Park business modes of "Base+E-commerce" and "Commerce and Trade + Logistics." Since 2012, the Linan Group has vigorously promoted and copied the "Linan Mode" in many cities across the country.

Up to the end of 2014, more than 20 commerce and trade logistics park projects had been successfully duplicated.

56PINGTAI is a logistics platform enterprise which, through its own informatized platform, adopted the "Internet+Logistics (park)" mode to perform the standardized upgrading/rebuilding and integration on traditional logistics parks to build the national chain of highway logistics parks. Up to the end of 2014, 56PINGTAI had integrated logistics parks with the overall area of 6.7 million m^2 in 28 logistics hub cities across the country in 18 months. The move had changed the prior situation of disperse operations and information isolated-island of traditional logistics park in China, with the high-quality standard management service, and markedly improved the operations efficiency of these logistics parks.

1.3.5.2 Joint Development of Construction Alliance of Dedicated Highway Transportation Enterprise

2014 was a year that dedicated highway alliance in China emerged. The dedicated highway enterprises in China are mainly small and medium-sized transport companies. In response to the depressed market demand and the rising operating cost, small and medium-sized special-line enterprises resorted to cooperating through the alliance platform, so as to expand the business, integrate the transport resources, increase the revenue, and build the brand name.

In January 2014, Huazhondadao Express Alliance was founded in Wuhan, the capital of Hubei Province in Central China. Capitalizing on the logistics brand name of "Wuhan Boulevard," the Alliance was founded by joining more than 10 high-quality small and medium-sized special-line enterprises. It plans to expand its membership to 200 in five years through the form of joint investment, share holding and entity operations. Presently, the Alliance operates 100 lines in Hubei, has 200 trans-provincial lines, with daily departure of 2,300 trucks; it has an annual transport volume of 25 million tons, accruing an annual revenue of 22 billion RMB. The Alliance has evolved into a national express network, transporting from the large and medium cities in Hubei directly to the nation's provincial capital cities, and radiating to the neighboring counties, cities, villages and towns.

In August, Heilongjiang Lanmeng Logistics Co., Ltd. was founded. Presently, it has 16 members units, more than 500 owned vehicles in the city, and more than 300 special-line distribution vehicles. In the Northeast region, it has more than 300 nodes and 128 branch lines; and its daily cargo throughput reached 6,000 tons, and the total warehouse area was 30,000 m^2.

In November, Wanzhong Logistics Alliance located in the eastern coastal city, Tianjin, was founded. This Alliance is composed of thirteen private logistics enterprises in Tianjin, and is the first logistics alliance run in the form of enterprise group in China. The operations warehouse area of the Alliance is 50,000 m^2, and the annual cargo throughput is more than 1 million tons. Based in Tianjin, it has 23 special lines shipping directly to nearly 100 first-tier cities to date.

1.3.5.3 Large-Scale Express Enterprises Strengthening Peer Cooperation and Service Expansion

In December 2013, four companies, i.e. YTO Express, STO Express, ZTO Express and YUNDA Express, which were among China's top six companies in express industry, built feng-wang.com. This move signifies that they have transformed from competition to competition/cooperation. Fengwang Company is mainly focused on four aspects: (1) invest in the upstream industry of the network retailing and express industry chains; (2) promote the adoption of intelligent self-service cabinet for express delivery to the community, campus and public transportation sites, using Fengwang as the platform; (3) promote the cross-border e-commerce of the four companies; (4) following Fengwang as the leader, jointly purchase various equipment and materials, including trunk line transport vehicles.

In 2014, the domestic express enterprises in China also accelerated the transformation towards comprehensive logistics enterprises. For example, SF Express started the convenience store strategy in 2014, built the community comprehensive service platform based on its self-run convenience stores. It acquired many enterprises which provide intra-city distribution services, so as to improve its "last-mile delivery" service capability. It then launched the "SF Cold Chain" brand by integrating the current resources in logistics, e-commerce, and convenience stores, to provide cold food warehousing, cold trunk line transportation, cold food home delivery, fresh food sales, supply chain finance and other one-stop services for clients in the fresh food industry.

1.3.5.4 China Railway Corporation Vigorously Developing New Freight Transport Products

In 2014, China Railway Corporation officially abolished the restriction on cargo types, and accepted all types of cargos, except for hazardous goods. National Railway Corporation also cooperated with e-commerce and express enterprises, and embarked on a series of new businesses, such as rapid freight delivery, e-commerce scheduled lines, and luggage and package delivery by high speed railway. Up to the end of 2014, 16 long-distance rapid freight delivery trains have been added in Beijing-Shanghai, Beijing-Guangzhou and other main trunk lines; six e-commerce express freight lines were opened; the luggage and package delivery by high speed railway has covered the main cities along the high-speed rail lines. About 500,000 parcels are delivered each day; daily shipment via the rapid freight delivery has risen to 60,000 tons from 10,000 tons in the initial stage, and has kept the increasing momentum.

In addition, China Railway Corporation has implemented the just-in-time transportation, thus, greatly compressed the arrival time window for freight. The system improved the cargo handling quality, reduced the freight loss and damage to the maximum extent, and lowered the freight cost. By continually perfecting the receiving and delivery network, and improving the door-to-door transportation service level, the railway logistics service functions are thoroughly improved. It has started to develop and manufacture high-speed rail freight trains, so as to make preparation for opening the e-commerce express line on the high-speed rail on a large scale.

1.4 Summary

This chapter introduces the economic and policy environment of China's logistics in 2014, the overall scale of its logistics market, and the main characteristics of the logistics market. For the economic environment, the world economy was recovering slowly, the global trade took on a sluggish growth, and China's economy has entered in the "new normal" development phase. The Internet economy in China was developing rampantly, while new business types and new business modes appeared in large numbers. For the policy environment, the Chinese Government actively advocated and promoted the "One Belt and One Road" initiative, and accelerated the development strategy of free trade zones, so that level of openness to the external world was continually raised. The development strategy of two major regions, i.e., the "Coordinated Development of the Beijing-Tianjin-Hebei Region" and the "Yangtze River Economic Belt," was promulgated, which signified that these two regions will become China's future development focus of regional economy.

For the overall scale of the logistics market, the growth rates of total social logistics, total social logistics expense, added value of logistics industry, and other indexes were slowed down to some extent. The growth of the freight volume and freight turnover was relatively steady; the cargo throughput of port and cargo throughput of airport showed an increased aggregate volume with a declined growth rate. The business volume of express delivery still maintained a high-speed growth.

For the main development characteristics of the logistics market, the cross-border logistics, urban joint distribution, and rural logistics system construction were developing promisingly. Logistics has drawn much attention from domestic and overseas investors, and many sub-fields have become the investment hot spots. Logistics enterprises continued to promote the transformation and upgrading of China's logistics industry by actively introducing the mobile internet and big data technology, implementing the chaining development, building the industry alliance, and developing the new logistics service products.

References

China e-Business Research Center (2015) Data monitoring report of China's e-commerce market in 2014. cnfol.com. http://news.cnfol.com/chanyejingji/20150409/20507776.shtml. 9 Apr 2015

iResearch Inc (2015) Transaction scale of online shopping market in China. iResearch.com. http://www.iresearch.com.cn/view/246308.html. 9 Feb 2015

Pintu360 (2015) O2O development report in catering industry in China in 2014. Pintu360.com.http://www.pintu360.com/article/54d701f414ec53c11661063e.html. 28 Jan 2015

Chapter 2
Logistics Facilities and Technological Development

Fan Qin

This chapter describes the progress in the development of transportation infrastructures, logistic nodes and channels in China in 2014; it also elaborates on the latest development of logistics facilities, logistics IT development and standardization. In 2014, guided and driven by the "One Belt and One Road" initiative and the national strategy of "Yangtze River Economic Belt," the scale of China's transportation infrastructures was further expanded; its comprehensive transportation network was improved, its layout of logistics parks (centers) became more rational, and logistics parks were evidently transformed and upgraded. Meanwhile, the technological level of logistics facilities was elevated, and the number of automatic and environmentally-friendly facilities kept increasing. Application of IT and Internet technologies was deepened in the field of logistics, and logistics standardization process was carried on orderly.

This chapter has four sections. Section 2.1 presents a general overview of the construction of China's transportation infrastructures and the development of its comprehensive transportation system. Section 2.2 analyzes the new characteristics in the development of China's logistics parks (centers) and warehousing facilities. Section 2.3 describes the changes in quantity and technological level of two categories of logistics facilities, i.e., logistics instruments and warehousing facilities. Section 2.4 discusses the development of China's logistics IT and the latest development in logistics standardization.

F. Qin (✉)
Business School, Nankai University, Tianjin, People's Republic of China
e-mail: fannq@163.com

L. Wang et al. (eds.), *Contemporary Logistics in China*, Current Chinese Economic Report Series, DOI 10.1007/978-981-10-1052-1_2

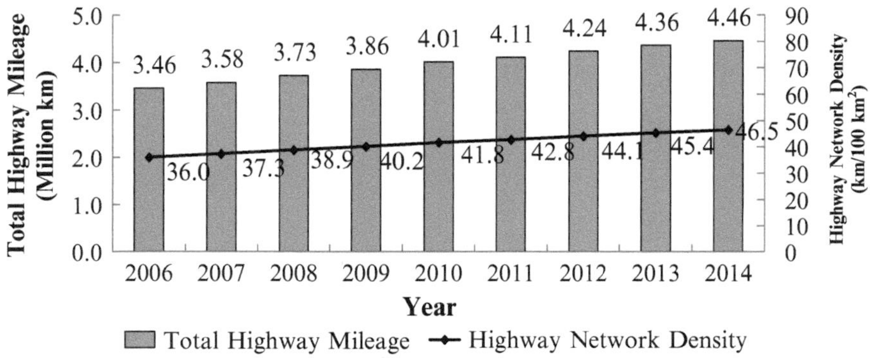

Fig. 2.1 Total highway mileage and highway network density for 2006–2014
Source: Compiled from the *National Statistical Yearbook for 2011–2014,* published by the National Bureau of Statistics, the *Highway and Waterway Transportation Industry Statistical Bulletin for 2006–2012,* and the *Transportation Industry Statistical Bulletin for 2013–2014,* published by the Ministry of Transport

2.1 Current Conditions in the Construction of Transportation Infrastructure in China

In 2014, the comprehensive transportation system in China continued to improve. Tangible growth was made in the scale of transportation network in Central China, Western China, and rural areas of the country. Structure of transportation network was further optimized. All these progresses provided sound transportation infrastructures to support the "One Belt and One Road" initiative and the "Yangtze River Economic Belt" strategy.

2.1.1 Highway Infrastructure

2.1.1.1 Scale of Highway Network

In 2014, China's investment in highway development amounted to 1,546 billion RMB, grown by 12.9 % over the previous year. By the end of 2014, total mileage of highways in China had reached 4.46 million km, up by 2.5 % from the previous year; highway density had reached 46.50 km/100 km^2, up by 2.5 % from the previous year (Ministry of Transport of P.R.C. 2015). Figure 2.1 shows the total highway mileage and highway network density for 2006–2014.

2.1.1.2 Structure of Highway Network

Structure of highway network in China continued to be optimized and the proportion of graded highways in the total highway mileage continued to rise in 2014. The mileage of graded highways in China grew by 145,300 km to a total mileage of 3.90 million km; this total is 87.4 % of the total highway mileage, and represents an increase of 1.2 percentage points over the previous year. The expressway mileage reached 111,900 km, growing by 7,500 km from the previous year; the mileage of grade-II and above highways reached 545,600 km, rising by 21,300 km, or 12.2 % of the total highway mileage, with the proportion gaining by 0.2 percentage points (Ministry of Transport of P.R.C. 2015).

2.1.1.3 Highway Construction in Rural Areas

In 2014, grading of China's rural highways continued to improve. China invested 303 billion RMB in developing rural highways, rising by 20.4 % year-on-year. Total mileage of rural highways reached 3.88 million km, increasing by 96,800 km from the end of the previous year. 98.08 % of towns (townships) and 91.76 % of administrative villages had access to hard-surfaced roads, rising by 0.28 and 2.76 percentage points, respectively, from the end of the previous year (Ministry of Transport of P.R.C. 2015).

2.1.1.4 Highway Construction in the Western Region

Highway development made new progress in Western China in 2014. Several expressways were put into operation in Western provinces of Guangxi, Gansu and Xinjiang, thus constituting various logistics channels on the Silk Road Economic Belt and those connecting Western China with Central and Eastern China. These constructions have provided necessary transportation support for advancing the Silk Road Economic Belt initiative.

2.1.2 Railway Infrastructure

2.1.2.1 Scale and Structure of Railway Network

Railway network in China continued to expand in 2014. The mileage of operating railways rose by 8.4 % from the previous year and reached 112,000 km, making 2014 the fastest-growing year for railways since 2006. The railway network density stood at 116.48 km/10,000 km^2, rising by 9.04 km/10,000 km^2 from the previous year (National Railway Administration of P.R.C. 2015).

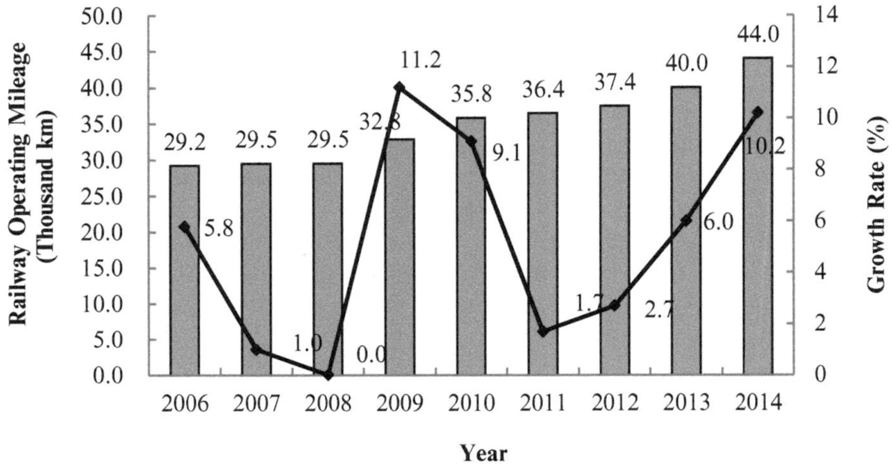

Fig. 2.2 Total railway operating mileage and growth rate of the Western Region for 2006–2014
Source: Compiled from data in the official website of the National Bureau of Statistics and the *Railway Statistical Bulletin 2014,* published by National Railway Administration of P.R.C

Railway network structure continued to be optimized in 2014. Mileage of operating railways in Western China continued to rise and reached 44,000 km, increasing by 4,020 km or 10.2 % from the previous year. Mileage of double-tracked railways rose considerably, up by 17.7 % from the previous year, amounting to 57,000 km; the proportion of double-tracked railways was 50.8 %, increasing by 4.0 percentage points from the previous year. The mileage of electrified railways grew by 16.9 %; the proportion of electrified railways reached 58.3 %, rising by 4.2 percentage points from the previous year (National Railway Administration of P.R.C. 2015). Figure 2.2 gives the total railway operating mileage and growth rate in Western China for 2006–2014.

2.1.2.2 International Railway

In 2014, new progress was made in international railway construction in China and connectivity with neighboring countries was improved. In February 2014, Tongjiang-Low Leninsquawe Railway Border Bridge, the first railway bridge crossing the river between Russia and China, commenced the construction. Upon completion, it will become another international transport channel that connects the railway network in Northeast China and the Trans-Siberian Railway in Russia, thus improving the existing layout of international railway transport in China. The Trans-Asia Railway connecting Southeastern Asia and China is the crucial carrier for implementing the "21st Century Maritime Silk Road" strategy. In December 2014, Mengzi-Hekou Railway, the last section in the eastern part of Trans-Asia Railway in China was put into operation, marking that Yunnan Province has established the first international railway channel connecting with Southeast Asia.

2.1.2.3 Railway Construction for Coal Transportation

In 2014, new breakthrough was made in the development coal-transporting railways and the improvement of its network structure in China. In December 2014, the heavy-load coal-transporting railway in Shanxi, Henan and Shandong was put into operation. It is another large-capacity coal-transporting channel in China, only after the Datong-Qinhuangdao Railway and the Shuozhou-Huanghua Railway that transport coal from Western China to Eastern China. This newly built railway can optimize the network structure and improve the ability of transporting coal out of the central and southern regions in Shanxi Province. Besides, the Gantang-Wuwei Railway II had commenced the construction. Upon completion, it will effectively alleviate the capacity bottleneck confronting the Baotou-Lanzhou, Lanzhou-Urumqi, Baoji-Zhongwei and Taiyuan-Zhongwei-Yinchuan railways, thus improving the energy transport in Northwest China.

2.1.2.4 High Speed Railway Construction in the Western Region

New progress was made in high-speed railway development in Western China. In December 2014, two major high-speed trunk railways in Western China, the Lanzhou-Urumqi and Nanning-Guangzhou high-speed railways, were put into operation. Lanzhou-Urumqi High-speed Railway is the first high-speed railway on the high-altitude plateau and desert in China; its operation has considerably shortened the transport distance and time between frontier regions in Northwest China and provinces in inland China. Meanwhile, the high-speed railway will greatly augment the railway transport capacity of Euro-Asia Continental Bridge, thus producing far-reaching influence on the New Silk Road Economic Belt. The operation of Nanning-Guangzhou High-speed Railway has improved the railway network structure in Central-Western China and Southern China.

2.1.3 Waterway Infrastructure

2.1.3.1 Scale of Waterway Infrastructure

In 2014, investment in waterway infrastructure was reduced somewhat in China. Investment expended to develop inland-river and coastal infrastructures was146 billion RMB, an amount contracting by 4.5 % from the previous year. Of this total investment, 50.81 billion RMB was expended on inland-river infrastructures, falling

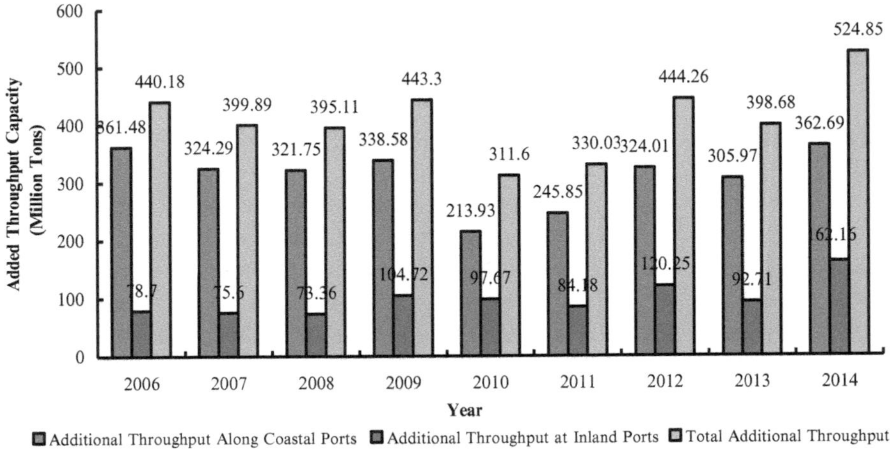

Fig. 2.3 China's added port throughput capacity for 2006–2014
Source: Compiled from the *Highway & Waterway Transportation Industry Statistical Bulletin for 2006–2012* and the *Transportation Industry Statistical Bulletin for 2013–2014*, published by the Ministry of Transport

by 6.9 % year-on-year, and 95.19 billion RMB was completed on coastal infrastructures, falling by 3.1 % year-on-year (Ministry of Transport of P.R.C. 2015).

Though total investment in waterway infrastructures was lower than that of previous years, the scale of construction continued to expand. For port construction, 170 berths were expanded or renovated in coastal ports and 363 million tons of throughput capacity was added; 253 berths were expanded or renovated in the inland-river ports and 162 million tons of throughput capacity was added. With respect to inland-river waterway construction, 2,000 km inland-river waterways were newly developed or improved. The waterway transport mileage in Yangtze River system and Pearl River system was increased considerably by 120 km and 281 km, respectively. By the end of 2014, the total operating mileage of inland rivers in China reached 126,300 km, up by 427 km from the previous year (Ministry of Transport of P.R.C. 2015). Figure 2.3 shows China's added port throughput capacity for 2006–2014.

2.1.3.2 Grading and Structure of Inland Waterway

In 2014, China's high-grade waterways in inland rivers developed rapidly; the proportion of high-grade waterways kept rising. By the end of 2014, the mileage of inland-river graded waterways in China reached 65,400 km, or 51.8 % of the total mileage of inland-river waterways, increasing by 0.2 percentage points from the previous year. The mileage of Grade-III and above waterways rose considerably and reached 10,850 km, or 8.6 % of the total mileage, rising by 0.5 percentage points from the end of last year (Ministry of Transport of P.R.C. 2015). Mileage of graded

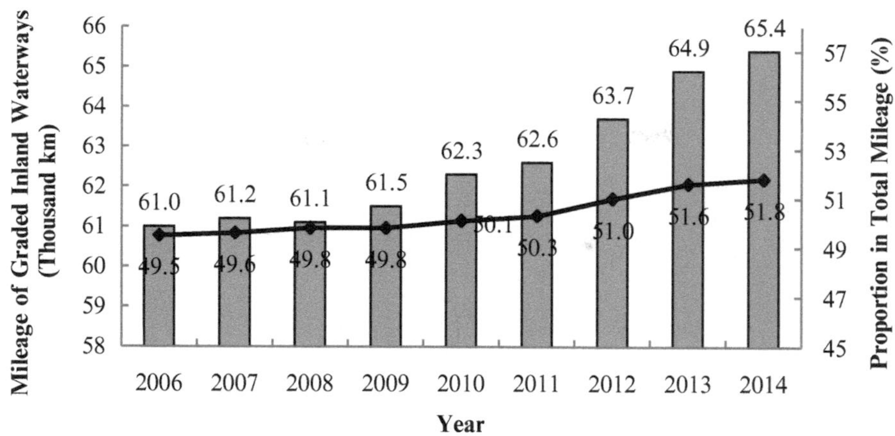

Fig. 2.4 Mileage of graded inland waterways and its proportion in total mileage for 2006–2014
Source: Compiled from the *Highway & Waterway Transportation Industry Statistical Bulletin for 2006–2012* and the *Transportation Industry Statistical Bulletin for 2013–2014*, published by the Ministry of Transport

inland waterways and its proportion in total inland waterway mileage for 2006–2014 are as shown in Fig. 2.4.

2.1.3.3 Berth at Port

In 2014, berths in China's ports continued to expand in size. By yearend, China's ports had 2,110 berths at or above 10,000 tons, increasing by 109 berths from the previous year. This total consists of 1,704 coastal berths at or above 10,000 tons, increasing by 97 berths with additional 331.23 million tons of throughput capacity; 406 inland-river berths at or above 10,000 tons, increasing by 12 berths with additional 30.94 million tons of throughput capacity. Figure 2.5 shows the number of port berths with capacity of 10,000 tons and above for 2006–2014.

2.1.4 *Civil Aviation Infrastructure*

In 2014, nine certified civil airports were newly built in China, six of which were in Central China and Western China. By yearend, China had 202 certified civil airports (excluding Hong Kong, Macao and Taiwan; same for below), including 200 airports with regularly-scheduled flights in 198 cities (Civil Aviation Administration of China 2015). Airport Distribution by Regions in 2014 is as shown in Fig. 2.6.

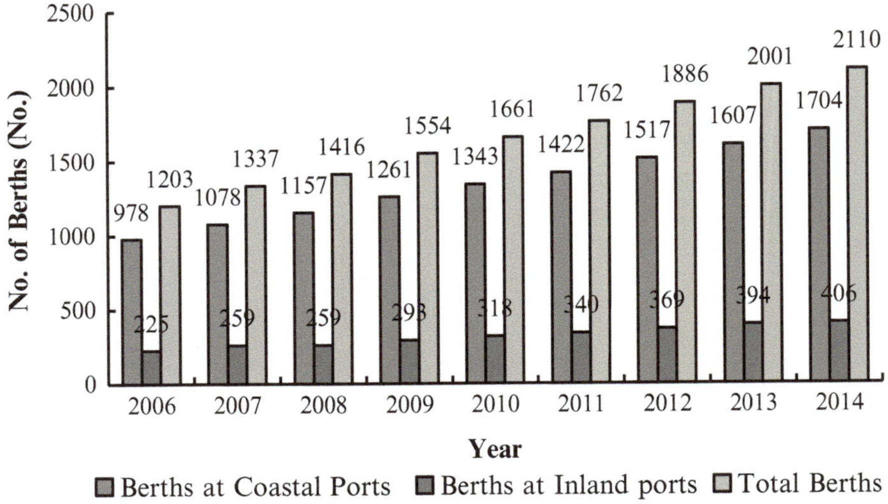

Fig. 2.5 Port berths with capacity of 10,000 tons and above for 2006–2014
Source: Compiled from the *Highway & Waterway Transportation Industry Statistical Bulletin for 2006–2012* and the *Transportation Industry Statistical Bulletin for 2013–2014*, published by the Ministry of Transport

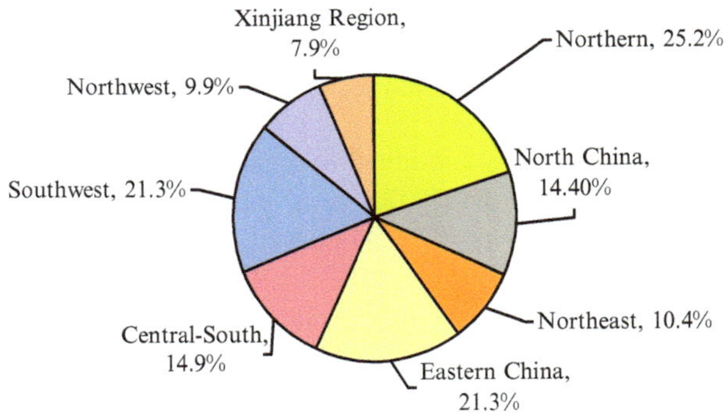

Fig. 2.6 Airport distribution by region in 2014
Source: Compiled from *the National Airport Productivity Statistical Bulletin 2014*, published by the Civil Aviation Administration of China (In the pie chart, the North China region includes Beijing, Tianjin, Hebei, Shaanxi and Inner Mongolia Autonomous Region; the Northeast China region includes Heilongjiang, Jilin and Liaoning; the East China region includes Shanghai, Jiangsu, Zhejiang, Shandong, Anhui, Fujian and Jiangxi; the Central-South China region includes Henan, Hubei, Hunan, Guangxi, Hainan and Guangdong; the Southwest China region includes Sichuan, Chongqing, Guizhou and Yunnan; the Northwest China region includes Shaanxi, Gansu, Ningxia and Qinghai; and the Xinjiang region refers to Xinjiang Autonomous region)

2.1.5 *Status of Comprehensive Transportation System*

2.1.5.1 Comprehensive Transportation Channel

In 2014, new progress was made in the comprehensive transportation channels in China, characterized by the "Five Vertical and Five Horizontal Lines" Schema. The logistics accessibility along the "One Belt and One Road" and in the "Yangtze River Economic Belt" was improved considerably. The grade of waterways along the Yangtze River, the major transport channel in Yangtze River Economic Belt, kept improving. Ships of 50,000-ton class could reach Nantong Port directly from the Yangtze River estuary; ships of 100,000 tons class could also reach Nantong Port during high tide. The waterways dredging project in Yangtze River's middle reaches was also going on smoothly. Besides, the Phase IV project of Yangshan Deep-water Port in Shanghai International Shipping Center commenced the construction; the Shanghai-Chengdu Expressway was put into operation.

New progress was made in developing the maritime transport channel in Northwest China and Northern China, including the Qingdao-Lhasa transport channel, the Manchuria-Hong Kong, Macao and Taiwan transport channel, the Baotou-Guangzhou transport channel, the Linhe of Inner Mongolia to Fangcheng Port of Guangxi transport channel.

Besides, the Weifang-Yingkou land-sea goods roll-roll, drop and pull transport channel had made the debut voyage. It was the first maritime green transport channel of its kind in China. Starting with Weifang Port and Yingkou Port, it covers Shandong and Liaoning as the central region, used the Bohai Bay waterways resources, and extends to the west bank of the Strait in the south and to Heilongjiang Province in the north. By making use of drop and pull transport depot system and land trunk network, the diverse transport channels were constructed, which ran for more than 1,000 km in the north-south direction in Eastern China (XinhuaNet 2014).

2.1.5.2 Construction of Joint Highway-Railway Bridge

In 2014, several joint highway-railway bridges were built or commenced, thus improving the logistics accessibility within some provinces, some regions, and even on the national scope. For instance, in March 2014, the "three-in-one" Shanghai-Nantong Yangtze River Bridge combining national railway, inter-city railway and expressway had commenced the construction. Also in June 2014, Hubei Huanggang Yangtze River Bridge, the railway-highway bridge with longest span in the world, was put into operation. The bridge has greatly improved the logistics accessibility in Central-South China.

2.2 Development of China's Logistics Parks (Centers) and Warehousing Facilities

In 2014, logistics parks (centers) showed evident signs of transformation and upgrading. Rapid progress was made in developing bonded logistics centers and commercial logistic parks in Western china; the network coordinated operating capacity among logistics centers started to improve; relevant plans of green warehousing were released, which afforded guidance for furthering the development of green warehousing facilities and management; construction of cold-warehouses continued to move on thanks to the stimulus from the rising demands of cold-chain logistics.

2.2.1 Development Status of Logistics Parks (Centers)

2.2.1.1 Medium and Long Term Development Plan for Logistics Parks Construction

In June 2014, the State Council passed the *Medium and Long Term Development Plan for Logistics Industry (2014–2020)*. The *Plan* again cited logistics park project as one of its twelve keystone projects. It clarified the requirements for developing logistics parks and stipulated the key issues in developing the logistics parks. Meantime, some provinces, autonomous regions and municipalities also issued plans, policies and measures to support the development of logistics parks. For example, Jiangsu Province in Eastern China and Sichuan Province in Western China announced plans of developing logistics parks in their respective regions. These plans will provide orderly construction of logistics parks in the respective regions.

2.2.1.2 Significant Trend Toward Logistics Parks' Transformation and Upgrading

In 2014, construction of logistics parks in China showed evident signs of transformation and upgrading. The development and operating models of logistics parks started to move from the preliminary phase of offering free or low-cost land plots to the new phase of emphasizing innovative services and management. In terms of revenue structure, in addition to the rent from warehouse/depot/office buildings, various value-added services have become the chief source of revenue for logistics parks, particularly in the well-developed coastal areas. More revenues came from industry integration, extension of industry chain and other value-added services. Logistics parks reflected the features of booming regional demands, definite functional positioning, strong awareness for service innovation, and apparent differentiation in operations (CFLP 2014a).

Meanwhile, the extent of network coordination in China's logistics parks was also enlarged gradually. As some logistics parks started to employ network operations, network logistics service systems based on such logistics parks and featuring rational layout, industrial cluster and functional integration were gradually established in some provinces and municipalities. These logistics parks were playing a better role in integrating regional network resources and stimulating coordinated development of regional economy. For instance, the Transfar's "Highway Port" in Zhejiang has built four highway ports in Hangzhou, Chengdu, Suzhou and Fuyang, and formed a logistics service network among these highway ports. Prologis Inc. invested in 35 major Chinese cities and built and managed 172 logistics parks. Basically, it has formed a logistics distribution network that covers major airports, sea ports, expressways, processing bases and some cities in China. Besides, enterprises operating in Linan Logistics Park and Blogis Logistics Park were also accelerating the pace of franchising to create the platforms for park resources, information and services.

2.2.1.3 Construction of Bonded Logistics Center

As the "One Belt and One Road" initiative and the "Yangtze River Economic Belt" strategy are implemented, China also sped up its construction of bonded logistics centers in 2014. Along the Yangtze River Economic Belt, Zhejiang Yiwu Bonded Logistics Center, and Bengbu Bonded Logistics Center were put into operation. Along the Silk Road Economic Belt, Xinjiang Kuitun Bonded Logistics Center, Chengdu Airport Bonded Logistics Center, and Wuwei Bonded Logistics Center, the first special Customs supervision region in Gansu, were put into operation. Besides, other regions also have many bonded logistics centers. For instance, Luzhong Yunda Bonded Logistics Center in Shandong was completed and put into operation in 2014.

2.2.1.4 Construction of Trade Logistics Parks in the Western Region

As e-commerce evolving rampantly and continuously in China, the new-type of international modern logistics parks, fueled by e-commerce, cross-border e-commerce, in particular commercial and trade logistics, developed rapidly. Many regions along the Silk Road Economic Belt (such as Sanyuan in Shaanxi, Lanzhou in Gansu, Yinchuan in Ningxia, Chongqing, Luzhou in Sichuan, Anning in Yunnan and Wuzhou in Guangxi) started to plan and build the new type of modern commercial and trade logistics parks, or to upgrade and renovate the existing commercial and trade logistics parks. This can conceivably help boost the regional logistics and regional economy of the local areas.

2.2.1.5 E-Commerce Logistics Network

As the logistics business of e-commerce enterprises continued to boom rapidly, the e-commerce enterprises also continued to expand and build their own logistics network in 2014. Dangdang.com and JD.com put into use their respective new large-capacity warehouse centers and logistics centers. Alibaba initiated the program of "Ten Thousand Villages in One Thousand Counties" to invest and build 1,000 county-level operating centers and 100,000 village service stations. In addition, many express transport, delivery and third-party logistics enterprises kept expanding their logistics networks to Central China, Western China, second-tier and third-tier cities and remote rural areas. According to statistics, nearly 50,000 express delivery outlets were newly developed in rural areas, and more than 2 billion parcels were delivered to rural areas in China (CFLP 2015).

2.2.2 *Development of Warehousing Facilities*

2.2.2.1 China's Green Warehousing and Distribution Action Plan

In April 2014, supported by the Ministry of Commerce, China Association of Warehousing and Storage released the "China's Green Warehousing and Distribution Action Plan" at the first China (International) Green Warehousing and Storage Conference. The Plan proposes using the new model and technologies of modern green warehousing and distribution, following the principles of "government guidance, market dominance, full alliance and systematic promotion" to facilitate the development of green warehousing and distribution in China in a holistic way, and make contributions to building intelligent and green commercial and trade logistic systems.

2.2.2.2 Area of Warehousing Land

In 2014, due to the impact of urban planning policies in China, land area for industries, mines and warehouses was reduced considerably to 150,000 ha, falling by 29.9 % from the previous year. Figure 2.7 gives the area of industrial, mining and warehousing land and growth rate for 2010–2014.[1]

[1] Since the land area for warehousing was not specified separately by the National Bureau of Statistics, the land area of industrial, mining and warehousing is cited here to give an indication of the overall change.

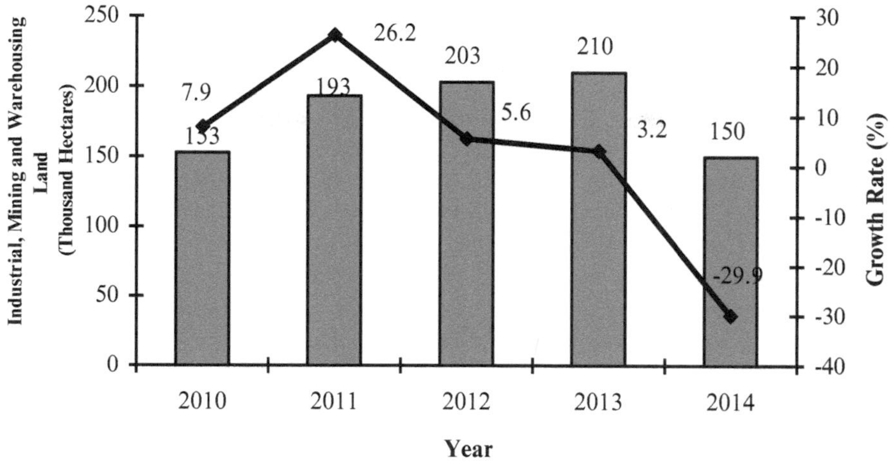

Fig. 2.7 Land area of industrial, mining and warehousing and growth rate for 2010–2014
Source: Compiled from the *National Economic and Social Development Statistics Bulletin for 2010–2014,* published by the National Bureau of Statistics

2.2.2.3 Rapid Growth in Construction of Cold Storage

As the commercial and trade logistics and agricultural product logistics developed, many refrigerated warehouses were built in 2014 to meet the increasing need for cold-chain logistics. According to the preliminary statistics of Cold Chain Logistics Committee of CFLP,[2] in 2014 China had 52 cold-warehouse projects at capacity of 10,000 tons and above under construction (Cold Chain Logistics Committee 2014). The Shenyang BGX International Agricultural Products Trading Center had cold-warehouse storage capacity of 400,000 tons; the Swire Pacific Cold Storage put into use its cold warehouses in Shanghai and Langfang, which are the largest cold warehouses ever put into operation in Eastern China, Beijing-Tianjin-Hebei region and, in Northern China. The Nanjing cold warehouse of Swire Pacific Cold Storage, currently under construction, has a capacity reaching 620,000 cubic meters.

2.3 Development of China's Logistics Equipment

In 2014, continual progress was made in the R&D of new-type of environmentally-friendly and heavy-load transport vehicles in China. Load capacities of various types of transport vehicles were increased. Enterprises, when selecting and utilizing warehousing equipment, were giving more attention to green, intelligent and standardized equipment so as to improve the logistics efficiency and realize sustainable development.

[2] CFLP refers to China Federation of Logistics & Purchasing.

2.3.1 *Development of Transportation Tools*

2.3.1.1 Possession of Transportation Tools

In 2014, the number of trucks in China continued to grow steadily to a total of 14.53 million units, rising by 2.4 % from the previous year. The number of aircrafts reached 2,365 units (Civil Aviation Administration of China 2014), up by 220 units from the previous year, making 2014 the fastest-growing year of aircrafts over the past five years. As cargo trains and heavy-load transport ships developed, China had 710,100 cargo trains in 2014, falling by 5,400 trains from the previous year. The number of transport ships stood at 172,000, falling by 0.3 % from the previous year (Ministry of Transport of P.R.C. 2015).

2.3.1.2 Technological Level of Transportation Tools

Loading capacity of various transport vehicles in China grew at different rates in 2014. The average tonnage of trucks reached 7.08 tons per unit, rising by 0.31 tons per unit from the previous year. The average net load of transport ships was 1,499 tons per ship, up by 6.0 % from the previous year; total number of containers reached 2.32 million TEUs, increasing by 36.3 % over the previous year, making 2014 the fastest-growing year for containers since 2006. In terms of railway transport, the China-made CRH5A high-speed train unit finalized the main-track test and will start the passenger test. Core technologies of high-speed railways and trains in China have achieved the capability of complete self-production.

2.3.2 *Development of Warehousing Facilities*

2.3.2.1 Construction of Nation-Wide Pallet Recycling and Pooling System

In 2014, the Ministry of Commerce conducted research on nation-wide pallet recycling and pooling system. After the system is put into practice, logistics cost can be reduced by at least 500 billion RMB per year. In terms of industrial practice, Shanghai Bailian Group Logistics Company directed its supermarkets to share the standard pallets with upper-echelon suppliers and to transport with holding plates, thus improving load/unload efficiency by four times and saving 18 % in number of pallets used. Taiyuan city focused on promoting the model of "intelligent label + IT-based management + socialized service" by utilizing standard pallets.

2.3.2.2 Usage of Environmentally-Friendly Forklifts

Along with the transformation of China's economy and the development of green logistics, Chinese enterprises accelerated the usage of electric forklifts in 2014. The demand for electric forklifts accounted for 30.6 % of the total demand for forklifts, rising by 3.5 % percentage points from the previous year, making 2014 the fastest-growing year over the past few years (56 Products Net 2015). In 2014, the sales volume of electric forklifts (including the electric balanced heavy-load riding fork-lifts and various electric warehousing forklifts) reached 110,200 units, rising by 23.92 % from the 88,900 units in the previous year (Jie Zhang 2015).

2.3.2.3 Expanded Usage of Conveyor and Sorting Equipment

Some enterprises has started to use the automatic sorting system, employing picking slips, computer bar code technology and RFID and other automatic sorting means, to accomplish the sorting operations more quickly and accurately, thus providing better service for clients. During the "November 11" promotion period in 2014, the number of express-delivered parcels reached 586 million pieces. Such an enormous number of parcels were delivered to the customers all over China thanks to the extensive use of automated sorting machines. In October 2014, the "Asia No. 1" Modern Logistics Center (phase I) of JD Mall in Shanghai was put into operation. The capacity of the center's sorting area reaches 16,000 pieces per hour, with a sorting accuracy of 99.99 %. The advance in automatic sorting technology has thoroughly resolved the problem of low efficiency and poor accuracy in manual sorting practices.

2.4 Development of Logistics Informationization and Standardization

IT development of logistics in China evolved rapidly in 2014. Due to progress made in the development of logistics cloud platforms at various levels, improvement in intelligent development of logistics parks, and swift promotion of Internet of Vehicles applications, the visibility level of logistics management and operations in China has been enhanced, and the logistics efficiency been improved as well. In terms of logistics standardization development, China set up a number of plans in 2014 on standardization of logistics, and a number of fundamental logistics standards and professional logistics standards.

2.4.1 Development of Logistics Informationization in China

2.4.1.1 Construction of National Public Logistics Information Platform

The national-level logistics public information platform refers to the nation-wide platform constructed by the Ministry/Commission of the Central Government or the national-level trade associations. It can, on a nation-wide arena, integrate logistics information resources, promote synergetic operations of logistics, and improve the comprehensive efficiency of logistics. In 2014, new progress was made in developing China's national-level logistics public information platform. In June 2014, China Logistics Financial Service Platform, co-supported by China Banking Association (CBA) and CFLP, was put online. By December 2014, the Ministry of Transport, the Ministry of Public Security, and the State Administration of Work Safety co-established the National Public Supervision and Service Platform for Highway Freight Vehicles. This Platform has more than 500,000 vehicles online, signifying that China has stepped up to a new informatization level in dynamically managing its highway freight vehicles. The Nation-wide Transportation Logistics Public Information Platform (LOGINK) has achieved the information connectivity among 18 ports of Korea, Japan and China in Northeast Asia; the Platform can provide information services in the unified platform standards, thus realizing connectivity of logistic information systems in these three countries.

2.4.1.2 Construction of Regional Logistics Cloud Platforms

In 2014, as the logistics IT development kept advancing, many new logistics cloud platforms emerged in China; these cloud platforms supported sharing and swapping of regional logistic information, and raised the efficiency in logistic operations, while providing constructive data bases for big-data type of logistics analysis. For instance, in July 2014, Gansu completed the building of its provincial-level "Beidou Logistics Cloud Integrated Service Platform;" enterprises such as China Post, JD.com, and INCITO had also launched their logistics cloud service in 2014.

2.4.1.3 Intelligence Level of Logistics Parks

The intelligence level of logistics parks in China was bettered in 2014. The new-type of O2O operations model, with logistics parks as the offline operators and public logistics information platform as the online coordinating tools, was promulgated gradually. For instance, the vehicle-goods matching platform for logistics parks has brought in the model used by taobao.com and DiDi Taxi to integrate sourcing of goods and vehicles; Transfar launched the "Intelligent Highway Port" model.[3] Development of IT platforms for logistics parks helps boost the intelligence level of logistics parks in China.

2.4.1.4 Application of Internet of Things (IOT)

In 2014, IOT technologies were used more extensively in China's logistics industry. The Ministry of Transport researched on the master plan of the new-generation national transport control network. It completed the "Intelligent Urban Transport" and the "Internet of Ships" national IOT demonstration projects, and the application of IOT for improving the monitoring of highway network operating status and efficiency. In particular, in the application of Internet of Vehicles, many IOV platforms were established in China, forming the closed eco-sphere of commercial IOV consisting of people, vehicles, roads, goods, networks and other logistics factors. These applications can actively push forward the intelligence level of regional logistics and logistic operations efficiency. By the end of 2014, the National Networking and Joint Controlling System for Key Operating Vehicles established by the Ministry of Transport had access to 31 provincial-level platforms; nearly 4 million vehicles were linked online. Besides, 56hyb.com, hdc56.com and other vehicle-goods matching App systems were gradually popularized in China (CFLP 2015).

2.4.2 Development of Logistics Standardization in China

2.4.2.1 The Medium and Long Term Development Plan for Logistics Standardization

In October 2014, the Standardization Administration of the People's Republic of China began to compile the *Medium and Long Term Development Plan for Logistics Standardization.* This document outlines the medium and long-term goals, major tasks and key fields in logistics standardization in China, as well as the special projects of logistics standardization in the next few years. The *Plan* will help perfect the framework of national logistics standards system and enhance the formulation of fundamental, public, service, and professional logistics standards. It is expected that by implementing the said *Plan*, a number of logistic standards will be established which will significantly promote the logistics development and service level in China (CFLP 2014b).

---

[3] Transfar's "Intelligent Highway Port" model refers to using IOT technologies, on the basis of traditional highway ports, to integrate the online information command system with offline logistics comprehensive supporting service system, thus realizing intelligent operations of highway port and improving the logistics efficiency.

2.4.2.2 Commissioning of the Working Platform for the National Logistics Standardization

In December 2014, the Working Platform for the National Logistics Standardization, which was co-established by the Department of Standardization Tasks of CFLP and the National Logistics Standardization Technical Committee, was put into trial operation. This Platform has set up the standards revision management system, the standards inquiry system, the online voting system, the standards questionnaire survey system, and the implementation feedback system; it is aimed to greatly propel the logistics standardization and elevate the service and management level.

2.4.2.3 Standardization Tasks for Trade Logistics

With the *Special Plan for Development of Trade Logistics* implemented and trade logistics developing rapidly, China released many plans on the development of trade logistics standards in 2014; these plans provided policy assurance for standardization in building modern trade logistics service system in China. In November 2014, the General Office of the Ministry of Commerce and the General Office of Standardization Administration jointly released the *Action Plan for Standardization of Trade Logistics.* This *Action Plan* addresses two major tasks of establishing the pallet sharing system and the comprehensive logistics information service platform (Department of Circulation Industry Development, Ministry of Commerce of the P.R.C. 2014).

2.4.2.4 Compilation of National Fundamental Logistics Standards

In 2014, according to the *12th Five-year Plan of Standardization Development* and the *Essence of 2014 National Standardization Tasks,* China formulated, revised, announced and implemented many fundamental logistics standards. These standards span various areas such as logistics services, operations, and IT development. Table 2.1 gives more details of these standards.

2.4.2.5 Compilation of National Professional Logistics Standards

In Recent years, thanks to the concerted efforts of the Vehicle Logistics Branch of China Federation of Logistics and Purchasing, some large motor manufacturers and logistic enterprises, a number of motor logistic standards have been constituted. In 2014, the four newly-formulated and implemented national standards have greatly helped resolve the problems of differed basic concepts in motor logistics, insufficient service process and indicators, and lacking in motor logistics service technical standards. Besides, they played an important role in regulating the motor logistics services, elevating the service quality, perfecting the development of motor logistics

Table 2.1 Formulated, revised, announced and implemented national fundamental logistics standards in 2014

Category	Name of standards	Status
National standards on logistics services	Classification and evaluation index for logistics enterprise	Implemented
	Specifications for logistics service contract	
	Service specifications and evaluation indicators for logistics parks	
	Statistical indicator system for logistics parks	
	Statistical indicator system of logistics prosperity index	
National standards on warehousing, cargo handling, transportation, packing	Technical specifications for container bulk cargo transport technical requirements for environmentally-friendly containers	Formulated
	Freight containers—coding, identification and marking	Revised
	Specifications and management requirements of stacking yards for refrigerated containers	
	Test methods for general-purpose flat pallets for through transit of goods	Announced
	Installation and maintenance specifications for automatic vertical warehouses	
	Fundamental elements and forms for warehouse receipt	Implemented
	Warehousing performance indicators system	
	Packing containers—heavy duty corrugated box	
National standards on logistics informationization	EbXML-based Logistic Information Reporting—request for Replenishing	Formulated
	EbXML-based Logistic Information Reporting—instruction and Notification for In-bound Warehousing	
	Standards on transport logistics information linking and sharing (2014)	Announced
	Numbering and bar-code Marking of Pallets	
	Swapping of transport logistics information–Part I: data elements swapping of transport logistics information—Part II: road transport electronic documents	Implemented

Source: Compiled from the 2014 *Notification Regarding the Revision of the First Batch of National Standards,* the 2014 *Notification Regarding the Revision of the Second Batch of National Standards,* and the 2014 *National Standards Bulletin No. 1–33,* by the Standardization Administrations of P.R.C.; the *Logistics Standardization Update, January-December 2014,* by the China Logistics and Purchasing Association and the *Category Handbook of Logistics Standards(2014), June 2014,* by the CFLP

standardization, and pushing forward a holistic development of China's logistics industry. In 2014, the international freight transport agency operations service standards were implemented, which helped regulate the international cargo agency market. Furthermore, the standardization of cold-chain logistics and hazardous goods logistics also made noted progress. The formulation, release and implementation of various professional logistics standards actively guided and regulated the logistics services, technologies and operations of several specific industrial sectors. Table 2.2 shows the status of these national professional logistics standards in 2014.

Table 2.2 Formulated, announced and implemented national professional logistics standards in 2014

Category	Name of standards	Status
Automobile logistics	Terminology of automobile logistics	Implemented
	Automotive logistics service evaluation index	
	Supervision requirements for damage risks of finished vehicle logistics	
	Automotive parts and components logistics—dimensions and technical requirements for plastic turnover box	
Trade logistics	Credit evaluation index system for trade logistics enterprise	Formulated
Cold chain logistics	Specifications for aquatic products cold-chain logistics service	Announced
	Service requirements and capability evaluation indicators of logistics enterprises	
	Safety requirements and test methods on refrigerated vans for transporting food and biological products on road	Implemented
	Specifications for air transport of cold-chain freight	
International logistics	Operational norms of sea-rail combined transportation for international freight forwarders	Implemented
	Operational norms for processing trade logistics	
Hazardous goods logistics	Land transport and storage requirements of fireworks and firecrackers	Formulated
	Methods of determining external safety distance for production and storage equipment of hazardous chemicals	
	Transport of hazardous goods—specification on the recognition and classification procedures and the requirement of compatibility for explosives	Implemented
	Specifications for storage and preservation of combustible and explosive goods	
Urban logistics	Specifications for building and operating urban distribution information platforms	Formulated
	Technical requirements for lectotype of urban logistics distribution vehicles	Implemented
Pharmaceutical logistics	Service specifications of medicine logistics	Implemented
	Technical requirements of infrastructures and facilities for medicine logistics	
Agriculture product logistics	Cotton baling	Implemented
	Technical criterion for grain and oil-seeds storage	

Source: Compiled from the 2014 *Notification Regarding the Revision of the First Batch of National Standards,* the 2014 *Notification Regarding the Revision of the Second Batch of National Standards,* and the 2014 *National Standards Bulletin No. 1–33,* by the Standardization Administrations of P.R.C.; the *Logistics Standardization Update, December 2013,* by the China Logistics and Purchasing Association and the *Category Handbook of Logistics Standards(2014), June 2014,* by CFLP

2.5 Summary

In 2014, the comprehensive transportation system in China, guided by the "One Belt and One Road" initiative and the national strategy of "Yangtze River Economic Belt," continued to move forward; the scale of transport network in Central China, Western China, and rural areas showed notable growth; the structure of national transport network continued to be optimized. China continued the push towards the construction, transformation, and upgrading of its logistics parks. In 2014, China built many bonded logistics centers, and constructed several trade logistics parks in the Western region, which promoted its foreign trade. Development of warehousing facilities moved towards green and intensive orientation. As cold-chain logistics developed, China facilitated building of cold warehouses with increasing number of large-sized ones. In terms of transport vehicles, the number of freight motors and airplanes in China kept the growing trend. However, as the loadings of transport vehicles became heavier, the number of freight trains and transport ships decreased from that in 2013. Warehousing equipment in China evolved towards intelligent and green development. In 2014, China activated several new national-level logistics public information platforms and regional logistics cloud platforms. As the logistics information App systems and logistics park information platforms were put into operation, the intelligence level of China's logistics parks was elevated. IOT technology, in particular the Internet of Vehicles technology, was more widely used in the logistics field. As various standardization work plans and national standardization working platforms were implemented, the standardization tasks in China's logistics industry will progress in a more regulated manner.

References

CFLP (2014a) Outline of Chinese logistics parks development, CFLP. http://www.chinawuliu.com.cn/wlyq/201405/30/290277.shtml. 30 May 2014

CFLP (2014b) Standardization administration started to compile the medium and long term plan for logistics standardization, CFLP. http://www.chinawuliu.com.cn/lhhkx/201410/24/294823.shtml. 24 Oct 2014

CFLP (2015) He liming: review of Chinese logistics industry development in 2014 and outlook for 2015, CFLP. http://www.chinawuliu.com.cn/lhhkx/201501/16/297616.shtml. 16 Jan 2015

Civil Aviation Administration of China (2014) 2015 national civil aviation work meeting & aviation safety work meeting were convened, Civil Aviation Administration of China. http://www.caacnews.com.cn/newsshow.aspx?idnews=262104. 26 Dec 2014

Civil Aviation Administration of China (2015) The national airport productivity statistical bulletin 2014. Civil Aviation Administration of China. 3 Apr 2015

Cold Chain Logistics Committee (2014) Part of the 10,000-ton+Cold Warehouses under Construction in China in 2014. Cold Chain Logistics Committee, 24 Jan 2014

Department of Circulation Industry Development, Ministry of Commerce of the P.R.C (2014) Special action plan for standardization of trade logistics, Department of Circulation Industry Development. Ministry of Commerce of the P.R.C. http://ltfzs.mofcom.gov.cn. 24 Nov 2014

Jie Zhang (2015) Overview of Chinese and overseas markets of industrial vehicles in 2014, vol 3. China Storage and Transport,http://www.chinachuyun.com/yuedu/zhuangbei/14235358352 2133.html

Ministry of Transport of P.R.C (2015) Transportation industry statistical bulletin 2014. Ministry of Transport of P.R.C. 30 Apr 2015

National Railway Administration of P.R.C (2015) Railway statistical bulletin 2014, National Railway Administration of P.R.C. 27 Apr 2015

Products Net (2015) Market prediction of major products such as logistic technologies and equipment in 2015, 56 Products Net. http://www.56products.com/News/2015-3-12/FDEF9GC2A 6363I5182.html. 12 Mar 2015

XinhuaNet (2014) Land-sea goods roll-roll, cross-docking transport channel in Shandong and Liaoning made the maiden trip on 23, Xinhua Net. http://www.sd.xinhuanet.com/wf/2014--12/24/c_1113753317.htm. 24 Dec 2014

Chapter 3
Policies and Plans Regarding Logistics Development in China

Ling Wang

While China is transforming and upgrading its economy, and adjusting its economic structure, the Chinese Government has been giving greater emphasis on the development of logistics. In 2014, the Government issued a number of policies to promote a holistic development of the logistics industry. In particular, the *Medium and Long Term Development Plan of Logistics Industry (2014–2020)* was issued, marking that logistics industry has been elevated to a national strategic industry. In 2015, the Government will focus on the major national strategies of the "One Belt and One Road," the "Yangtze River Economic Belt," and the "Beijing-Tianjin-Hebei Coordinated Development." China will issue more measures to promote development of the logistics industry, from the perspectives of building regional logistics channels, opening-up of the logistics industry, and logistics innovation based on the Internet.

This chapter contains three sections. Section 3.1 gives a general introduction to the policies regarding China's logistics development in 2014. This section outlines the policies issued in 2014, covering aspects such as the medium and long term planning, opening-up of the logistics industry, regional logistics coordination, transportation reform, transformation and upgrading of logistics industry, and the supervision of logistics market. Section 3.2 interprets *the Medium and Long Term Development Plan of Logistics Industry;* it describes the background and the major contents and characteristics of the *Plan*. Section 3.3 presents an outlook on the policies and plans regarding logistics in China. This section analyzes the development trend of China's logistics policies in five aspects, i.e., regional logistics channels, international logistics policies, synergistic development of logistics and relevant industries, Internet plus-based logistics innovations, and establishment of regional Customs clearance.

L. Wang (✉)
Logistics Research Center, Nankai University, Tianjin, People's Republic of China
e-mail: wlnk3006@163.com

L. Wang et al. (eds.), *Contemporary Logistics in China*, Current Chinese Economic Report Series, DOI 10.1007/978-981-10-1052-1_3

3.1 Status of Logistics Policies and Plans Issued in 2014

In 2014, China issued a number of policies to promote the holistic development of its logistics industry, in order to speed up the opening-up of the logistics industry, further the coordinated development of regional logistics, hasten the transportation reform, accelerate the logistics transformation and upgrading, and enhance the supervision of logistics market., The issuance of the *Medium and Long Term Development Plan of Logistics Industry (2014–2020)* signifies that China will pick up the pace of transforming and upgrading of its logistics industry to adapt to the "New Norm" of the nation's economy.

3.1.1 The Medium and Long Term Development Plan of Logistics Industry, the First of Its Kind for China

In November 2014, the State Council issued the *Medium and Long Term Development Plan of Logistics Industry (2014–2020)*. According to the *Plan*, by 2020, China will basically have built up a modern logistics service system featuring rational layout, advanced technologies, efficiency, convenience, greenness and environmental-friendliness, safety and order. To realize this objective, the *Plan* outlines three key points, i.e., reducing the logistics cost, improving the level of scale and intensity of logistics enterprises, and enhancing the establishment of logistics infrastructure and network. It also proposes seven major tasks such as improving the commercial and professional level of logistics, and twelve major projects such as multi-modal transport. The *Medium and Long Term Development Plan of Logistics Industry (2014–2020)* is China's first synoptic document addressing the medium and long term development of logistics industry. It is of great significance in steering the transformation and upgrading of the logistics industry facing the new economic situation. Section 3.2 examines the *Plan in* more detail.

To implement the *Plan*, the National Development and Reform Commission (NDRC) and relevant departments issued the *Three-year Action Plan to Promote Development of Logistics Industry (2014–2016)* in December 2014. The *Action Plan* proposed 62 major tasks in five aspects, and designated the spearheading department, the specific targets, and the completion deadline of each task. Major tasks in the *Thee-year Action Plan to Promote Development of Logistics Industry (2014–2016)* are shown in Table 3.1.

Table 3.1 Major tasks in the *Three-year Action Plan to Promote Development of Logistics Industry (2014–2016)*

No.	Targets	Major contents
1	Reducing the logistics cost	Streamline administration, delegate power to the lower levels, and deepen the reform of administrative approval systems
		Rectify and cleanup the practice of unauthorized fees and fines on highway transport
		Build the grand logistics channels
		Improve the operation and management of urban distribution vehicles
		Implement the tax- supporting policies
2	Improving the level of scale and intensity of logistics enterprises	Encourage the logistics enterprises to develop trans-regional network operations
		Actively cultivate large-sized logistics enterprises
		Implement and improve the policies on land usage that support the development of logistics enterprises
		Expand the investment and financing channels for the logistics enterprises
		Encourage the logistics industry to open up and go abroad
3	Enhancing the establishment of logistics infrastructure and network	Speed up the construction of facilities for multi-modal transport
		Fortify the demonstrative and leading role of logistics parks
4	Promoting the formation of key logistics projects	Push forward grain warehousing logistics and building of agricultural product cold-chain logistics facilities
		Enhance building logistics facilities for resource-type products
		Improve the urban and rural distribution network
		Develop e-commerce logistics and speed up the establishment of public logistics information platforms
		Promote the standardization of logistics
		Push forward the development of green logistics
		Improve the emergency-response logistics system
5	Performing fundamental work	Improve the laws, rules and regulations, regulate the market order
		Enhance the safety supervision, statistical analysis, and talent cultivation

3.1.2 Opening-Up of Logistics Industry Relying on the Development of Free Trade Zone

To deepen the reform and widen the opening-up to the outside world, in December 2014, the State Council announced the establishment of China (Guangdong) Pilot-test Free Trade Zone, China (Tianjin) Pilot-test Free Trade Zone, and China (Fujian) Pilot-test Free Trade Zone, while expanding the scope of Shanghai Free Trade Zone. These three new FTZs were the first batch approved in China after the State Council granted approval to China (Shanghai) Pilot-test Free Trade Zone in August 2013.

Modern logistics is a major content in the building of FTZs. Tianjin Free Trade Zone will build North China International Shipping Center and International Logistics Center while pilot-testing policy innovations in four aspects, i.e., international ship registration system, international shipping taxation, shipping finance business, and leasing business. Zhuhai Port and Yantian Port in Guangdong Free Trade Zone will represent the frontier development of port logistics industry. Zhuhai Port is targeted at "building the first-class port logistics operator and domestically-renowned comprehensive energy investment service provider." Fujian will make use of the FTZ to bolster the contacts with Taiwan. By mobilizing the relevant parties in and out of the FTZ, it will explore the Taiwan-oriented logistics model and build the strategic channel connecting Mainland China and Taiwan. Shanghai will leverage the opportunity from the expansion of its FTZ to actively transform from the commodity trade to the service trade, and establish the new framework of international logistics trade. As these four FTZs develop, logistics industry will, at a higher level, support China to open wider to the outside world. Meanwhile, the FTZs will become the frontier of the opening-up of China's logistics industry.

3.1.3 *Coordinated Development of Regional Logistics*

Regional logistics is an essential part of regional economy. In 2014, the State Council, the Ministry of Commerce and other departments issued various policies regarding the planning of regional economic development; all of which included regional logistics development in their planning so as to play the supporting role of regional logistics, and promote coordinated development of regional economy. The Yangtze River Economic Belt and Northeast Region will receive major development efforts.

3.1.3.1 Building of Comprehensive and Vertical Transport Corridor in the Yangtze River Economic Belt

The Yangtze River waterways, as the Golden Waterways with largest inland-river freight volume in the world, is the most important east-west axis in the overall layout of regional development in China. However, the Yangtze River comprehensive transport system is yet to be better developed; presently the Golden Waterways fail to robustly stimulate the economic growth of the vast hinterland. In September 2014, the State Council issued the *Guiding Opinion of the State Council about Relying on the Golden Waterways and Pushing Forward Development in the Yangtze River Economic Belt,* which aims to advance the Golden Waterways system and speed up the development of the Yangtze River Economic Belt.

The *Opinion* raises requirements for the development of logistics industry in two aspects. First, optimize the functional layout of ports, strive to develop modern shipping industry, and form a number of regional logistics shipping centers. Meanwhile, enhance building the collection/dispersion/transport system, develop multi-modal transport, and push forward building channels between ports and logistics parks. Second, enhance the strategic interaction with the Silk Road Economic Belt, play Chongqing city's central and hub role in the west of the Yangtze River Economic Belt, fortify the strategic support to the Silk Road Economic Belt, so as to form the clustering effect of regional logistics and build the modern international comprehensive transport hub.

To support the *Opinion*, the State Council, meanwhile, issued the *Plan of Comprehensive and Vertical Transport Corridor on the Yangtze River Economic Belt (2014–2020), which* focuses on three aspects. First, build up the Yangtze River Golden Waterways. This requires designing major projects on the trunk and branch waterways along Yangtze River and building of the collection/dispersion/transport system. Second, build the comprehensive transport corridor. The *Plan* requires enhancing the railway transport network, optimizing the highway transport network, expanding the aviation transport network, setting up the comprehensive transport nodes, and building the international logistics channels. Third, speed up the construction of transport network for city clusters, so as to establish the transport networks for four major city clusters, namely, the Yangtze River Delta city cluster, the Yangtze River Middle Reach city cluster, the Chengdu-Chongqing city cluster, and the city cluster of Central Guizhou and Central Yunnan. The implementation of the *Plan* will help fulfill the potential of Yangtze River waterways, improve the transport capacity of east-west highways and railways, and enhance the connection among the infrastructures. While carrying out the *Plan*, China will uncover the potential of domestic demand in the hinterland along the upper reaches of Yangtze River, extend the space of economic growth from the coastal areas to the inland areas along the Yangtze River, optimize the industrial structure and urbanization layout of cities along the river, and upgrade the economy for the Yangtze River Economic Belt.

3.1.3.2 Synergistic Development of Logistics in Northeast China

To speed up the transformation and upgrading in Northeast China and other old industrial bases, China released the *Few Opinions of CPC Central Committee and the State Council about Implementing the Strategy of Rejuvenating Northeast China and Other Old Industrial Bases* in 2003. In 2009, the *Opinions of the State Council about Implementing the Old Industrial Base Rejuvenation Strategy* was released. At present, considerable progress has been made in the economic and social development of Northeast China. However, the region is still facing new challenges, e.g., continuing slowdown of economic growth rate, difficulties in production and operations for some industries, and glaring and in-depth institutional, systemic and structural problems. In August 2014, the *Opinions of Major Policies and Measures of the State Council on Supporting Rejuvenation in Northeast China* was issued, pointing out that it was imperative to consolidate and expand the results of prior rejuvenating efforts, and endeavor to find measures for resolving problems with the development. In Particular, the *Opinions* raised the following requirements for developing the area's logistics and transportation.

First, enhance the building of grain storage and logistics facilities. It proposed building a number of standardized storage facilities for grain, enhancing construction of transport channels for grain, improving the grain logistic system and layout of nodes. Second, build a number of major platforms for open cooperation. In regions with proper conditions, build the comprehensive bonded area and cross-border economic cooperation area, and speed up building the Dalian Northeast Asia International Shipping Center. Third, speed up the building of comprehensive transport networks. Hasten the building of express railways within the three provinces in Northeast China; promote the building and extension of airports in Changchun, Harbin, Shenyang and Dalian, finalize the construction of China-Europe railway channel via Russia, and accelerate the construction of China-Mongolia railway channel.

Implementation of the above *Opinions* will, on the one hand, improve the grain logistics system and layout of nodes through enhancing the construction of grain-transporting channels and logistics infrastructures, hence facilitate the channel of transporting grains from North China to South China. On the other hand, by building the international transport channels and logistics nodes, the crucial position of Northeast Region in China's international logistics system will be highlighted.

3.1.4 Overall Deepened Reform in the Field of Transportation

In 2014, the General Office of the State Council, the National Development and Reform Commission and the Ministry of Transport released a number of documents on deepening the reforms in the field of transportation, which aimed at furthering the reform in certain major fields and key sections.

3.1.4.1 Guiding Opinions on Deepening the Transportation Reforms

In December 2014, the Ministry of Transport issued the *Opinions of the Ministry of Transport on Deepening Comprehensive Transportation Reforms*, which was designed to promote the modernization of the transportation governance system and governance capability. According to the *Opinions*, by 2020, China will have established the transportation system that supports and assures the national economic, social development and better livelihood of the people. Its major contents are: (1) improve the comprehensive transportation institution and mechanism and the modern market system of transportation; (2) speed up the transformation of government functions, deepen the reform of administrative approval system, enhance the responsibilities of public-service and market supervision; (3) deepen the reform in transportation investment and financing institution and the reform of highway/waterway management system; (4) improve the modern transportation service system.

The *Opinions*, as the steering document set to guide the comprehensive and in-depth reform in the transportation industry till 2020, will help China attain major breakthroughs in the fields such as macro transportation, investment and financing institution/mechanism, and toll highways.

3.1.4.2 Speeding Up Reforms in Investment and Financing Institutions of Transportation Infrastructures

To implement the *Opinions of the State Council about Reforms in Railway Investment and Financing Institutions and Speed up Railway Construction* (State Council Document [2013] 33), and carry out the policies of comprehensive development and the utilization of railway land and the areas neighboring depots in supporting railway development, the General Office of the State Council, in August 2014, issued the *Opinions of the General Office of the State Council about Supporting the Railway Construction and Implementing Comprehensive Land Development.* The latter *Opinions* advocates liquidizing the existing land used by railways, and adopts the policies of comprehensive use and development of railway land and the areas neighboring depots. These policies are major measures to attract the non-government capital and speed up the reforms of railway investment and financing institutions, and railway construction. They are helpful in alleviating the fund pressure on railways and accelerating the development of railway infrastructures.

In May 2014, the NDRC issued the *Notification about Releasing the First Batch of Projects Encouraging Investment in Infrastructures and Other Fields,* and resolved to launch the first batch of 80 demonstration projects of attracting non-government capitals to participate in the construction and operations of infrastructures. For infrastructures of railway, highway and port, the *Notification* encourages and attracts the non-government capital, in particular, the private funds, to participate in the construction and operations, in the form of joint venture, wholly-owned investment and franchised operation. These actions help speeding up the investment and financing institutional reform in transportation infrastructures, promoting diversification of investors, ensuring the non-government capital playing its role in the construction of infrastructures, and alleviating the fiscal pressure brought on by the infrastructure construction and operations for local governments.

3.1.4.3 Market-Oriented Reform in the Pricing of Transport Services

To ensure the market plays the deciding role in resource allocation, China has improved the fare policies in railway, aviation, and port; specifically, China has delegated some pricing authorities on transportation fares and liberated some of the fees for transportation services.

In November 2014, the Ministry of Transport and the NDRC issued the *Notification of A Few Issues about Liberating the Fares for Port's Competitive*

Services. The *Notification* requires liberating the port's fee standards on labor and ship services, regulating the calculation methods for labor fee, simplifying the items for fee collection, enhancing supervision over the port's fee collection behaviors, and improving the port's fee collection regulations. In the same month, the Civil Aviation Administration of China (CAAC) and the NDRC issued the *Notification about Improving Civil Aviation Transportation Price Policies,* which proposed liberating the price of transporting freight on domestic civil aviation routes, and implementing the clearly-posted pricing practice. In December 2014, the NDRC issued the *Notification of Empowering and Liberating Prices of Some Transportation Services,* which dictated the delegation of pricing authority of short-distance channel transport within the province., i.e., price of short-distance channel transport within the province to be controlled by the provincial-level pricing administration. In the same month, the NDRC issued the *Notification about Liberating Prices of Some Railway Transport Products,* which proposed that railway transport enterprises can determine on their own the price of express transportation for bulk cargos and regular transportation of parcels on railways.

Pricing reform is a crucial content in pushing forward the market-oriented reforms and the transformation of government functions. Liberating the service fees in the transportation field can ensure the market playing a deciding role in setting prices, stimulate the dynamics of market players, establish the market adjustment system, and promote positive operations in the transportation market.

3.1.5 Speeding Up Transformation and Upgrading in Port Shipping and Logistics Industry

To alleviate pressure on the shipping market, the State Council and the Ministry of Transport released several documents to push forward the transformation, upgrading and healthy development of shipping industry in 2013. In the following year, relevant departments issued more policies to steer the development of waterway transportation industry.

In June 2014, the Ministry of Transport issued the *Guiding Opinions of the Ministry of Transport on Pushing forward Port Transformation and Upgrading.* According to the *Opinions*, China would focus on altering the development modes of the ports, and establish, by 2020, the modern port service system featuring quality and benefits, greater hub functions, greenness, safety, intensity, efficiency, and convenience. The *Opinions* spelled out six major tasks in port transformation and upgrading: (1) expand service functions and develop modern port industry; (2) improve the port's transportation system and push forward building of comprehensive transportation hub; (3) allocate port resources on a rational basis and steer towards intensive port development; (4) enhance technological and management innovations and pursue development of green ports; (5) strengthen safety management of ports and fortify construction safety of ports; (6) improve port informatization level and promote efficient and convenient port services.

In August 2014, the State Council issued the *Certain Opinions of the State Council on Promoting Holistic Development of Ocean Transport Industry*, which pointed out that, by 2020, China would build the modern ocean transport system with international competitiveness featuring safety, convenience, efficiency, economy, and greenness; China would expect its ocean transport service trade to rank in the leading position in the world while markedly improving the international competitiveness. To realize these goals, the *Opinions* raised seven major tasks: (1) optimize the structure of shipping fleet; (2) improve the global shipping network; (3) promote the transformation and upgrading of ocean transport enterprises; (4) vigorously develop the modern shipping service industry; (5) deepen the reform and opening-up of ocean transport industry; (6) upgrade the international competitiveness of ocean transport industry; and (7) push forward the safe and green development. The *Opinions* is China's first national top-level document on the development of ocean transport, and the first holistic and systematic plan on the development of ocean transport, marking that developing ocean transport has been elevated to a national strategy.

To implement the *Certain Opinions of the State Council on Promoting Holistic Development of Ocean Transport Industry,* the Ministry of Transport, in December 2014, issued the *Opinions of the Ministry of Transport about Speeding up Development of Modern Shipping Service Industry,* which pointed out that, by 2020, China would have basically established the modern system of shipping service industry featuring full functions, quality service, efficiency, convenience, and orderly competition. Pursuant to the overall goal, the *Opinions* outlines major tasks in eleven aspects: (1) promote the transformation and upgrading of traditional shipping service industry; (2) improve the capability of shipping transaction service; (3) innovate in shipping finance and insurance service; (4) enhance the capability in marine legal service; (5) improve the capability in shipping information service; (6) enhance the price indexing service function; (7) enhance service in ship technology; (8) improve the crew's service capability; (9) improve the market supervision system of modern shipping service industry; (10) deepen the international cooperation and exchange; (11) improve the service function of shipping centers. The *Opinions* serves as the detailed version of the *Certain Opinions on Promoting Holistic Development of the Ocean Transport Industry.* It is also the guideline document for the development of shipping industry in China during the 13th Five-year Plan period, which helps promote China's transformation and upgrading of its shipping industry, and raising the industry's international competitiveness.

3.1.6 Endeavoring to Improve Professional Level in the Logistics Fields

In a clear-cut way, the *Medium and Long Term Development Plan of Logistics Industry (2014–2020)* specifies "improving the professional level of logistics" as one of the three major tasks. In 2014, China issued a number of policies in the fields of commercial and trade circulation, agricultural products and cold-chain logistics, and urban delivery, so as to guide and encourage the logistics fields to evolve toward the professional orientation.

3.1.6.1 Commercial and Trade Logistics

In September 2014, the Ministry of Commerce released the *Opinions on Promoting the Development of Commercial and Trade Logistics,* which outlined the tasks in the development of commercial and trade logistics as follows. (1) Improve the level of commercial development. Support the commercial and trade logistics enterprises to enhance their supply-chain management, strive to develop the joint distribution, support the transformation and upgrading of traditional warehousing enterprises, and develop towards distribution operations centers and third-party logistics. (2) Raise the level of professionalization. Endeavor to develop e-commerce logistics, enhance the development of cold-chain logistics, speed up the transformation and upgrading of production-materials logistics, and encourage the development of green logistics. (3) Improve the level of standardization. Nurture innovations in mechanisms of disseminating, implementing and promoting standards. Begin with pilot test of cycled and pooled used of standard pallets and gradually improve the penetration rate of standard pallets in the society. (4) Improve the level of informatization. Support the enterprises' shared information system; realize data sharing, resource sharing and information connectivity, and support the development of comprehensive logistics information service platforms pertaining to enterprises. (5) Improve the level of organization. Encourage the logistics enterprises to grow stronger and bigger and pursue collaborative development, guide the joint investment to build warehousing facilities in major logistics nodes, and carry out the rational layout of logistics parks (centers and bases). (6) Improve the level of international development. Promote the building of international logistics channels, support the building of overseas economic and trade cooperation zones featuring commercial and trade logistics, and encourage the capable commercial and trade logistics enterprises to go international and conduct global business.

3.1.6.2 Agricultural Products and Cold-chain Logistics

To solve the problems in the Chinese agricultural product market such as fragile system, high cost of distribution, and low-efficiency distribution, thirteen departments in China including the Ministry of Commerce, jointly issued the *Guiding Opinions about Enhancing the Development of Agricultural Products Market System* in February 2014. According to the *Opinions*, China would invest 5–10 years to successively establish an efficient, smooth, safe and regulated agricultural-product market system with Chinese attributes and orderly competition. The *Opinions* pointed out that China would, in major distribution nodes and regions with advantages in agricultural products, speed up building a number of agricultural-product distribution centers, pricing centers, processing and distribution logistics centers and international exhibition and sales centers of agricultural products, which would have domestic and international influence. It also proposed to support the construction of facilities for producing on-site pre-cooling, preliminary processing and storage of agricultural products, to support distribution enterprises to integrate resources in upstream production and downstream marketing, to promote coordinated development of agricultural-product cold chain, supply chain, IOT and Internet, so as to improve the cold-chain circulation rate of agricultural products.

In December 2014, ten departments including the NDRC jointly issued the *Guideline Opinions on Promoting holistic Development of Cold-chain Transport Logistics Enterprises*, which gave the following directives: (1) by way of cold-chain transport and logistics enterprises mutually sharing and holding equities, improve the scale and intensity level of cold-chain transport; (2) enhance the construction of cold-chain logistics infrastructures; (3) improve the standardized system of cold-chain transport and logistics; (4) actively push forward the informatization of cold-chain transport and logistics; (5) vigorously develop the joint distribution and other advanced models of organizing distribution; (6) optimize the permit administration measures for urban distribution vehicles; (7) enhance and improve the industry supervision; (8) provide greater supporting policies in taxation; and (9) effect the proper functions of trade associations. The *Guideline Opinions* is the synoptic documents issued in this period regarding cold-chain transport and logistics, which underscores the Government's apparent attention and support to cold-chain transport and logistics.

3.1.6.3 Urban Distribution

In recent years, delivery vehicles in Chinese cities have been confronted with difficult problems in "accessing, parking, and loading/unloading." In February 2014, the Ministry of Transport, the Ministry of Public Security, and the Ministry of Commerce jointly issued the *Notification of Enhancing the Management of Urban Distribution Transport and Vehicle Traffic*, which proposed eight measures in enhancing the urban distribution transport and vehicle travel management: (1) enhance the demand management for urban distribution capacity; (2) strengthen the technological management of urban distribution vehicles; (3) regulate the development of urban freight service vehicles; (4) optimize the traffic management measures for urban distribution vehicles; (5) improve the management measures for parking of urban distribution vehicles; (6) improve the service level of urban distribution transport; (7) enhance the supervision and management of urban distribution transport market; (8) improve the management mechanism of urban distribution transport and vehicle traffic. The *Opinions* would play an important role in regulating the operations and improving the traffic environment of urban distribution transport, and alleviating the traffic congestion in the cities.

3.1.7 Supervision and Regulated Development in the Logistics Market

3.1.7.1 Supervision of Cross-Border E-Commerce

In August 2013, the General Office of the State Council transmitted the *Notification of Implementing Relevant Policies on Supporting Retail and Export in the Cross-border E-commerce* issued by the Ministry of Commerce. This document is the so-called "States' Six Articles" by the Chinese Government in supporting the development of cross-border e-commerce industry; it is China's first policy regarding cross-border e-commerce development.

In December 2013, the Ministry of Finance and the State Administration of Taxation jointly issued the *Notification of Retail and Export Taxation Policies in Cross-border E-commerce*, which clarified the preferential taxation policies regarding the retail and export of cross-border e-commerce. In January 2014, the General Administration of Customs issued the *Notice of Adding Code in Customs Supervision Measures (No. 2014[12])*, which enclosed cross-border e-commerce into the Customs supervision with the "9610" code, whose full name is "cross-border trade e-commerce." In July of the same year, the General Administration of Customs released the *Notice of Adding Code in Customs Supervision Measures (No. 2014[57])*, which added the Customs supervision measures code "1210," whose full name is "bonded cross-border trade e-commerce." In August 2014, the General Administration of Customs issued the *Notice of Supervision of Entering & Exiting Goods and Articles in the Cross-border E-commerce,* which would supervise enterprises and individuals engaging in the cross-border business, so that goods purchased overseas would be governed under relevant regulations.

3.1.7.2 Building of Logistics Credit System

In China, the logistics industry is currently poorly organized and the market players are conspicuously "small, scattered and disorderly." Some enterprises develop or operate in a poorly-regulated manner, resulting in frequent violations of laws or regulations and defaulting. In November 2014, spearheaded by the NDRC, the Ministry of Transport, the Ministry of Commerce, the National Railway Administration, the CAAC, the State Post Bureau and the Standardization Administration jointly issued the *Guideline Opinions on Building the Credit System for the Logistics Industry in China*, which proposed the following measures. (1) Endeavor to foster and support the credit service organizations. (2) Vigorously push forward building of credit report and require relevant departments to improve the credit information collection mechanism. (3) Ensure that credit records on logistics industry are widely utilized in the society as a whole. (4) Build the credit incentive and default punishment mechanism. (5) Establish and improve the laws, rules, regulations and standards regarding logistics credit. (6) Enhance the building of corporate honesty system, and guide logistics enterprises to cultivate the philosophy of honest operations. (7) Strive to push forward the establishment of industry honesty and credit culture. (8) Vigorously push forward the building of honesty in administrative offices. (9) Conduct pilot test in credit building in professional logistics fields, such as cold-chain logistics, hazardous products logistics, motor logistics, and other logistics fields that have developed more maturely.

The *Opinions* is the first guideline document in China about building honesty and credit system in the logistics industry. Its issuance is of significant implications in regulating the operations of enterprises and building the market environment featuring fair competition and honest operations.

3.2 Interpretation of the Medium and Long Term Development Plan of Logistics Industry

The State Council issued the *Medium and Long Term Development Plan of Logistics Industry (2014–2020)* in 2014. Prior to this release, the State Council had issued the *Plan of Readjusting and Rejuvenating the Logistics Industry* in April 2009, and the General Office of the State Council had issued the *Opinions on Policies and Measures of Promoting Holistic Development of Logistics Industry* in 2011. The above latest document sums up issues in the development of logistics industry and provides an overall and synoptic development plan.

3.2.1 Background of Issuing the Plan

In recent years, logistics industry in China has seen growing rather rapidly; its service capability improved noticeably, its infrastructures and policy environment upgraded considerably, and its modern industrial system basically established. Logistics industry has become an integral part of the national economy. However, the overall development level of the logistics industry is not very high, and the development approach is quite coarse. First, the logistics cost is high but the efficiency is low. Second, the segregation within the industry is severe and the mechanism and institutional barriers hindering industrial development have not been broken. Third, infrastructures are lagging behind relatively, which cannot meet the requirements of modern logistics development. Fourth, the system of instituting policies, laws and regulations is deficient, and the market order is not well-regulated.

In July 2011, the NDRC led other thirteen departments to establish the Steering Panel for Compilation of the *Medium and Long Term Development Plan of Logistics Industry (2014–2020)*. On June 11, 2014, Premier Li Keqiang chaired the Executive Meeting of the State Council to pass the *Medium and Long Term Development Plan of Logistics Industry (2014–2020)*.

3.2.2 Major Contents of the Development Plan

The major contents of the *Medium and Long Term Development Plan of Logistics Industry (2014–2020)* are tabulated in Table 3.2 as follows.

Table 3.2 Major contents of the *Medium and Long Term Development Plan of Logistics Industry (2014–2020)*

Policy orientation	Key themes	Major contents
Development goal		By 2020, basically build the modern logistics service system featuring rational layout, advanced technologies, convenience, efficiency, greenness, environmentally-friendliness, safety and orderliness
Three key themes		Reduce the logistics cost
		Elevate the scale and intensity level of logistics enterprises
		Enhance the establishment of logistics infrastructure network
Seven major tasks	Improve commercial and professional level of logistics	Encourage manufacturing enterprises to outsource logistics business, and direct their internal logistics needs to go commercial. Encourage logistics enterprises to integrate functions and engage in business innovations, improve professional service quality and actively develop customized logistics service
	Enhance logistics Infomatization development	Enhance application of advanced IT such as Beidou navigation system, IOT, cloud computing, big data, and mobile Internet in the logistics field, and speed up building the enterprises' logistics information system
	Push forward modernization of technological equipment in logistics	Speed up the R&D of professional logistics equipment for food cold chain, pharmacy, tobacco, machinery, motor, dry and bulk goods, and hazardous chemicals, and improve the professional level of logistics equipment
	Strengthen the establishment of logistics standardization	Strive to compile and implement the medium and long-term plan of logistics standards, perfect the logistics standard system and improve the national framework of logistics standard system
	Coordinated development of regional logistics	As required by the Silk Road Economic Belt, the Maritime Silk Road and the Yangtze River Economic Belt strategy, speed up building of logistics channels in key logistics regions and those connecting China Mainland and other countries/regions
	Promote development of international logistics	Enhance building of port logistics infrastructures for key seaports, airports, railways and highways. Develop the border trade, spearheaded by developing and opening-up of pilot-test regions
	Vigorously develop green logistics	Optimize the transport structure, rationally allocate various transport means, improve the proportion of railway and waterway transport, and promote energy saving and emission reduction

(continued)

Table 3.2 (continued)

Policy orientation	Key themes	Major contents
Twelve major projects	Project of multi-modal transport logistics	Develop marine-railway transport, railway-waterway transport, highway-railway transport, and land-air transport; push forward the waterway-railway transport of bulk commodities, and multi-modal transport of containers
	Project of logistics parks	Develop logistics parks featuring freight hub type, production service type, commerce and trade service type, port service type, and comprehensive service type, as well as professional logistics parks of agricultural products, iron and steel, pharmacy, and express delivery
	Project of agricultural-product logistics	Resolve to build modern grain logistics facilities; enhance building of logistics nodes and special railways; strengthen the construction of "transporting sugar from South China to North China" and cold-chain logistics facilities of fresh agricultural products
	Project of manufacturing logistics and supply-chain management	Build the warehousing distribution facilities platform and logistics information platform; encourage industrial clusters and functional regions to build public external warehouses; bring in third-party logistics enterprises
	Project of resource type of products logistics	Push forward building of key coal channels in Shanxi, Shaanxi, Inner Mongolia (West), Ningxia, Gansu, East Inner Mongolia and Xinjiang; focus on building the large coal storage bases and key coal logistics nodes in circum-Bohai region
	Project of urban & rural logistics distribution	Improve the urban and rural distribution network system: make overall planning and rational layout of distribution nodes; build the urban distribution public service platforms; push forward building of distribution network system of consumer goods and agricultural materials in counties, townships and villages
	Project of e-commerce logistics	Build the regional warehousing &and distribution bases and attract the manufacturers, e-commerce businesses, express delivery and LTL logistics companies, and third-party service companies to the bases; explore the development of high-speed railway express delivery. Improve a batch of express-delivery transit centers by collaborating with the cross-border e-commerce pilot test
	Project of logistics standardization	Focus on pushing the revision of logistics standards in logistics technologies, information, service, transport, agricultural products, pharmacy, motor, e-commerce, post (including express delivery), cold-chain, and emergency response

(continued)

Table 3.2 (continued)

Policy orientation	Key themes	Major contents
	Project of logistics information platform	Integrate the existing resources of logistics information service platforms; build the intelligent logistics information public service platform. Enhance building of platforms for comprehensive transportation information, transaction of logistics resources, electronic port, and commodity trading; promote connectivity and information sharing among various platforms
	Project of developing and application of new logistics technologies	Push forward application of Beidou navigation system, IOT, cloud computing, big data, mobile Internet and other technologies in product traceability, online dispatching and management, automatic logistics distribution and intelligent distribution
	Project of recycling renewable resources	Focus on pushing forward development of recycle logistics on wastes from daily lives, such as packages, used electronic and electric products, and other wastes with residual value such as scrapped machinery, discarded raw materials produced in consumer goods production process
	Project of emergency-response logistics	Build the emergency-response logistics system featuring unified coordination, rapid response, orderly operation, reliability and safety; establish a number of pillar logistics enterprises with relative strong emergency-response logistics operation capability

3.2.3 *Major Attributes of the Development Plan*

3.2.3.1 Clarification of the Strategic Position of the Logistics Industry

The *Medium and Long Term Development Plan of Logistics Industry (2014–2020)* points out clearly that logistics industry "is the fundamental and strategic industry that supports the development of the national economy. Speeding up the development of modern logistics industry is of significant impact in promoting the adjustment of industrial structure, shifting the development orientation, improving the competitiveness of the national economy and building the ecological civilization." Hence, the status of logistics industry has been elevated, and its significance highlighted, signifying that China has gained a deeper understanding of the logistics industry.

3.2.3.2 The Holistic and Specific Development Targets

The *Plan* specifies the overall targets for the building and development of logistics industry in the upcoming five years. The development targets cover four aspects, i.e., logistics services, enterprises, infrastructures, and operating efficiency; they are holistic and specific, encompassing both quantitative indicators and qualitative ones. The targets in these four aspects are described briefly below.

1. Further elevating the commercial and professional level of logistics. The target is to increase the annual logistics industry's added value by about 8 %, and for the added value to reach about 7.5 % of the total GDP in China. Moreover, it will markedly raise the proportion of third-party logistics, and broaden the employment of new logistics equipment and technologies.
2. Evidently strengthening the competitiveness of logistics enterprises. The target is to improve the capability of integrated operations and network management, to notably raise the informatization and supply-chain management level, thus establishing a number of large groups of comprehensive logistics enterprises with international competitiveness and logistics service brand names.
3. Smoothly linking the logistics infrastructures and the operating modes. The target is to cause the layout of logistics park network system more rational, ensure the modern logistics operating means, such as multi-modal transport, drop and pull transport, and joint distribution, to develop more rapidly, thus highlighting the effect of clustered logistics development.
4. Clearly lifting the overall operating efficiency of logistics. The target is to lower the proportion of total cost of logistics in the GDP from 18 % in 2013 to about 16 %, and cause the logistics industry to better support and bolster the national economy.

3.2.3.3 Focus on Conspicuous Problems

The *Plan* stresses that the development of logistics industry should have three focal points. First, reduce the logistics cost; second, improve the scale and intensity level of logistics enterprises; third, enhance the logistics infrastructure network. These three focal points pinpoint the critical problems which constrain the development of the logistics industry and the breach which allows advancement in the development of the logistics industry.

3.2.3.4 Major Tasks and Key Projects as Handles

The *Plan* outlines seven major tasks in three aspects. The first aspect addresses the quality and effectiveness of the logistics industry; it proposes to elevate the level of socialization and professionalization of the logistics industry, to enhance the construction of logistics informatization, to push forward the modernization of logistics technologies and equipment, and to strengthen the establishment of logistics standardization. The second aspect focuses on the regional coordinated development of logistics industry; it proposes to progress towards the development of regional logistics and international logistics. The third aspect attends to the environmental protection; it proposes to vigorously develop green logistics. The *Plan* has clearly specified goals and tasks, and has strong targeting and guiding features.

The *Plan* outlines twelve key projects in the following five categories. (1) Infrastructures: project of multi-modal transport logistics, project of logistics parks; (2) Product-type logistics: project of agricultural-product logistics, project of manufacturing logistics and supply-chain management, project of resource-type products; (3) Operating modes: project of urban-rural logistics distribution, e-commerce logistics; (4) Fundamental work in the industry: project of logistics standards, logistics information platform, development and application of new logistics technologies; (5) Social responsibility: project of recycling renewable resources, and emergency-response logistics. These key projects involve the vital links and fields in the development of logistics industry; they represent the crucial measures which relevant departments and local governments may adopt in supporting the development of logistics industry.

3.2.3.5 Emphasis on the Realization of the Plan

The crux of the *Plan* is about implementation. The *Plan* outlines nine measures to ensure the implementation, including the deepening of reform and opening-up, improving laws and regulations, regulating the market order, enhancing safety supervision, improving the supporting policies, expanding channels of investment and financing, strengthening statistical work, promoting theoretical study and talent training, and leveraging the role of trade associations. To implement the *Plan*, the NDRC and relevant departments issued the *Three-year Action Plan to Promote Development of Logistics Industry (2014–2016)*, which specified the leading department, the concrete goals and deadlines for completing the tasks.

3.3 Outlook on China's Logistics Policies and Plans

At present, the Chinese economy is in the state of the "New Norm." The logistics industry also needs to deal with new problems arising out of the development through transformation and upgrading. In 2015, China will focus on the nation's primary strategies, namely, the "One Belt and One Road Initiative," the "Yangtze River Economic Belt" and the "Beijing-Tianjin-Hebei Coordinated Development," and adopt the relevant plans and policies, such as the *Medium and Long Term Development Plan of Logistics Industry (2014–2020)* and the *Three-year Action Plan to Promote Development of Logistics Industry (2014–2016),* while making new measures that encourage the development of the logistics industry.

3.3.1 More Funding Support for Building Regional Logistics Grand Channels

In 2015, the One Belt and One Road Initiative will be gradually implemented. In March 2015, the NDRC, the Ministry of Foreign Affairs, and the Ministry of Commerce jointly issued the *Vision and Action of Pushing Forward Co-building the Silk Road Economic Belt and the 21st Century Maritime Silk Road*. According to the *Vision*, China will endeavor to improve the regional infrastructures and form the safe and efficient land-sea-air channel network, so as to attain a new level of connectivity. The connectivity of infrastructures is the priority field in the One Belt and One Road Initiative. At present, the Silk Road Foundation, the AIIB, the BRICs Development Bank, and the SCO Development Bank are expected to provide financial support for the relevant projects in 2015.

The *Plan Outline of Beijing-Tianjin-Hebei Coordinated Development* will also be issued in 2015. The *Outline*, in addition to clarifying the positioning of these three province/municipalities, will also establish the target for coordinated development in the region. Transportation integration will be the leading field in the Beijing-Tianjin-Hebei Coordinated Development. Besides, according to the *Plan of Comprehensive Vertical Transport Corridor in Yangtze River Economic Belt (2014–2020)*, establishment of the comprehensive multi-dimensional transport corridor in Yangtze River region will be accelerated.

3.3.2 International Logistics Policies Based on the Building of FTZs Will Be More Open

In March 2015, the Master Plan of Pilot-test FTZs in Guangdong, Tianjin and Fujian and the *Plan of Deepening Reforms and Opening-up of Shanghai Pilot-test FTZ* were approved.[1] In April, the three FTZs in Guangdong, Tianjin and Fujian were put into operation, signifying that China has entered a new stage in building the FTZs.

Guangdong FTZ, by making full use of its location attributes, will focus on pushing forward the liberalization of service trade in Guangdong, Hong Kong and Macau; the logistics service in these three regions will identify new models of market connectivity, funds interchange, and mutual service recognition. As the important port city in North China, Tianjin FTZ will fully learn from the experience of Shanghai pilot-test FTZ, highlight the shipping business, and build specialties such as shipping taxation and finance. Fujian FTZ, based between China Mainland and Taiwan, will expand the opening-up of service trade across the Taiwan Strait and enhance the cooperation with Taiwan in service areas such as telecommunication

[1]The Master Plan of Pilot-test FTZs in Guangdong, Tianjin and Fujian was approved and policy welfare gradually got released. http://finance.people.com.cn/money/n/2015/0324/c42877-26743777.html

and transportation, commerce and trade, construction industry, product certification, engineering technologies, and professional technologies.

3.3.3 Linked Development of Industries Will Be Enhanced

The *Medium and Long Term Development Plan of Logistics Industry (2014–2020)* clearly specifies the professionalization and socialization development of logistics as its major tasks. In May 2014, the State Council Executive Meeting resolved to speed up the development of the critical as well as frail points in the production service industry, so as to promote the adjustment and upgrading of industrial structure. In this process, China emphasized on promoting linked development of the third-party logistics industry and the manufacturing industry. In 2015, China proposed the "China Manufacturing 2025" strategy to push forward the upgrading of manufacturing industry. The Ministry of Industry and Information Technology also lists intelligent manufacturing as its main task. The transformation and upgrading of manufacturing industry has raised new requirements for the development of logistics industry. In the future, logistics industry in China will not only deepen the integrated and linked development with the manufacturing industry, but also expand the scope of industries so linked.

3.3.4 Logistics Innovation Based on "Internet +" Will Be Encouraged

The 2015 Chinese Government Work Report pointed out, for the first time, to "set up the "Internet +" Action Plan and push forward mobile Internet, cloud computing, big data and IOT to bond with modern manufacturing industry; to promote the holistic development of e-commerce, industrial Internet and Internet finance."

In the backdrop of "Internet +," the transformation and upgrading of traditional logistics will embrace new opportunities. In April 2015, the National Transport Logistics Public Information Platform and Transfar jointly promoted the Park Express, the first Internet product delineated as the national logistics park information steward; this marks the initial integration of information platform resources. In the future, China will migrate from basically establishing logistics information platforms to encouraging networked and flattened development of logistics information platforms, and fostering a number of logistics information platforms which have market recognition and can show forth the demonstration effect. On the other hand, as China encourages "Mass Entrepreneurship and Mass Innovation," the Internet-based logistics innovation models will get stronger policy support, if they provide logistics service oriented towards O2O and vehicle-goods matching APP.

3.3.5 *Cross-Regional and Inter-departmental Grand Customs Clearance Will Be Hastened*

In February 2015, the State Council released the *Notification of the State Council about Issuing and Implementing the Plan of Building and Reforming the Grand Customs Clearance Featuring "Three Mutuals."*[2] *The Notification* advocated *mutual* information sharing across departments and regions, *mutual* recognition of supervision, and *mutual* assistance in law enforcement in port management. The target is to build the collaboration mechanism and management mechanism of grand Customs, and improve the grand Customs clearance management mechanism by 2020.

In the future, China will shift from the development of "Single Window" practice in international trade and the "serial law enforcement" to the "parallel law enforcement." In addition, it will develop the grand Customs clearance in various aspects such as the one-stop clearance operation, the mechanism of information sharing, in optimizing law enforcement resources of ports, innovations in consolidated law enforcement pilot study, and improving the opening-up arrangement of ports.

3.4 Summary

This chapter synthesizes China's logistics policies and plans in 2014, and presents an outlook on the trend of policies and plans for 2015 and onward. As China's economy has entered the phase of the New Norm, the Chinese Government is paying greater attention to the logistics development by issuing a number of synoptic, steering, and constructive policies and plans, thus pushing forward the transformation and upgrading of logistics in a holistic way. In particular, the issuance of the *Medium and Long Term Development Plan of Logistics Industry (2014–2020)* signifies that logistics industry has become a national strategic industry. This *Plan* is China's first guideline document on the medium and long-term development of logistics industry; it addresses China's logistics development in multiple levels and all-encompassing perspectives by setting development goals, major tasks and essential implementation projects. As the logistics industry steps up to a higher level, the Government will continue to issue logistics-related policies and promote the transformation and upgrading of the logistics industry, by focusing on building regional logistics channels, speeding up the opening-up of logistics industry to the outside world, pushing forward efficient and close industrial collaboration, encouraging the innovation in "Internet +" logistics models, and building the grand Customs clearance systems.

[2] General Administration of Customs interprets the Plans of Building and Reforming the Grand Customs Clearance Featuring 3 Mutuals. http://fangtannew.customs.gov.cn/tabid/335/InterviewID/65/Default.aspx.

Chapter 4
Development of Regional Logistics Along the One Belt and One Road

Jun Liu

The "One Belt and One Road" initiative is a major Chinese strategy of diplomacy and economic development. From the global perspective, the One Belt and One Road connects Europe, Asia, Pacific and Indian Ocean. The strategy will help establish the new international pattern of opening-up. From China's domestic perspective, the One Belt and One Road opens up the passages to the South and West of China, while promoting cooperation of flows of materials, business and information among regions in China; it benefits the coupled development of the Eastern, Central and Western regions of China. As such, the One Belt and One Road is also an important strategy for the logistic development in China. It works as the important means for the country to build the land and maritime logistic passages, and to establish and develop the two-way international logistic system.

This chapter contains four sections. Section 4.1 introduces the background of advocating the One Belt and One Road initiative and the major contents of the strategy. Section 4.2 elaborates on the economic development of major countries along the One Belt and One Road and their economic and trade relations with China. Section 4.3 describes the development of logistic infrastructures in those countries along the One Belt and One Road, and Section 4.4 presents an outlook on the regional logistic development along the One Belt and One Road.

J. Liu (✉)
Logistics Research Center, Nankai University, Tianjin, People's Republic of China
e-mail: lj369369@sina.com

L. Wang et al. (eds.), *Contemporary Logistics in China*, Current Chinese Economic Report Series, DOI 10.1007/978-981-10-1052-1_4

4.1 Overview of the One Belt and One Road Initiative

The One Belt and One Road initiative is a development strategy proposed by the Chinese leader. In a new fashion, it will forge a closer connection between and countries in Asia, Europe and Africa, thus linking up the development strategies, and promoting the economic prosperity and cooperation among the countries along the One Belt and One Road.

4.1.1 *Connotations of the One Belt and One Road*

In 2013, the Chinese President Xi Jinping announced the proposals of building the Silk Road Economic Belt and the 21st Century Maritime Silk Road (abbreviated as "One Belt and One Road"). The One Belt refers to a belt-shaped region of economic cooperation. With the cultural emblem of the ancient Silk Road, it includes China and other countries along the ancient Silk Road in Central Asia, Southeast Asia, South Asia, West Asia and European countries. The One Road refers to the strategic maritime passage connecting East Asia, South Asia, Southeast Asia, Africa and Europe. Figure 4.1 exhibits the detailed routes of the One Belt (marked red) and One Road (marked blue).

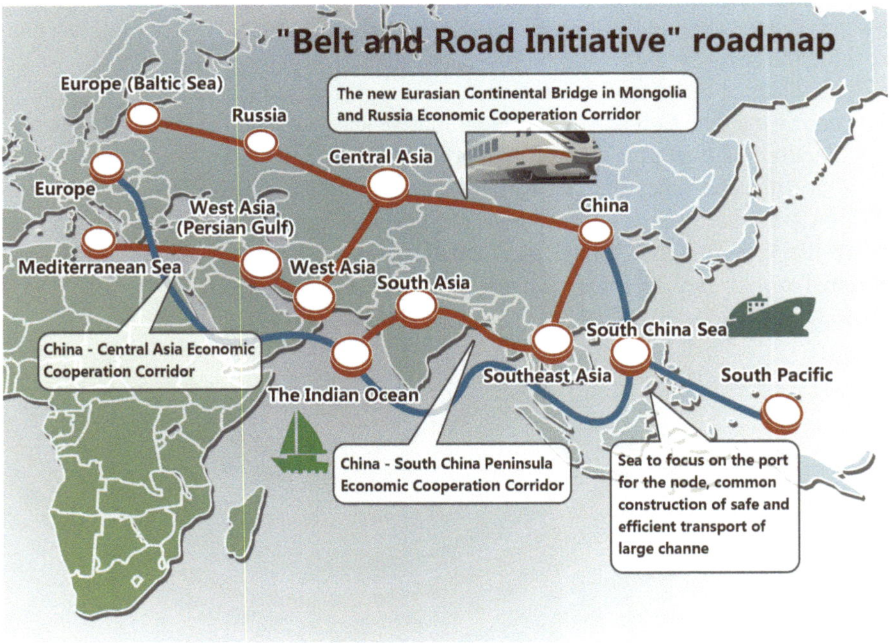

Fig. 4.1 Map of one belt and one road
Source: According to the schematic of One Belt and One Road, released by CCTV http://news.cntv.cn/2015/04/13/VIDE1428923854189562.shtml

The Silk Road Economic Belt has three directions. The Northern Route starts with China, goes through Central Asia and Russia, and reaches Europe. The Central Route starts with China and goes through Central Asia, West Asia, reaches Persian Gulf and the Mediterranean. The Southern Route starts with China, goes through Southeast Asia, South Asia, and reaches the Indian Ocean. Inside China Mainland, the One Belt spreads over to major regions such as Shaanxi, Gansu, Ningxia, Qinghai, Xinjiang, the three provinces in Northeast China, Inner Mongolia, Guangxi and Yunnan. Of these regions, Xinjiang functions as an important transport hub and center of commerce, trade, logistics, culture, science and education along the Silk Road Economic Belt. It is also the core of the Silk Road Economic Belt. From the international perspective, the One Belt extends to major countries and regions including the five countries in Central Asia, Russia, Indo-China Peninsula,[1] India, Pakistan, Iran and some European countries. The five countries in Central Asia connect the East Asia Economic Circle on one side and the Europe Economic Circle on the other side. It is the key region that helps realize the smooth development along the Silk Road Economic Belt.

The 21st Century Maritime Silk Road has two directions. The first route starts with the coastal ports in East China, goes through South China Sea, Indian Ocean, and reaches Europe. The second route starts with the coastal ports in East China and goes through South China Sea and reaches South Pacific. The One Road focuses on promoting the economic corridors such as the China-Pakistan Economic Corridor and the Bangladesh-China-India-Myanmar Economic Corridor, while developing an economic belt of strategic cooperation facing the South China Sea, Pacific Ocean, and Indian Ocean.

4.1.2 Background of Proposing the One Belt and One Road Initiative

The One Belt and One Road initiative was proposed with the view of the domestic and international background having panoramic and far-reaching significance. Internationally, TPP and TTIP have levied great pressure on China's foreign trade, pushing China to develop new market space. Domestically, there still exists considerable disparity of development between Central/Western China and Eastern China. Proceeding toward the West and the South helps solve the problems of imbalanced economic development and energy security.

[1] Indo-China Peninsula is one of three major peninsulas in South Asia. It is named after its location to the south of China and includes seven countries, i.e., Vietnam, Laos, Cambodia, Myanmar, Thailand, Singapore and Malaysia.

4.1.2.1 Incessant World-Wide Enhancement of Regional Economic Cooperation

Nowadays, regional trade agreement has become the new means of regional cooperation in the world. Such agreements, by offering more preferential trade and investment conditions, link the economic benefits of member states more closely while enhancing their political, diplomatic and cultural connections. Because of the lesser scope and lower trading cost, regional agreements have embraced prosperous development[2] after the WTO Doha Round Talks halted. TPP and TTIP are two regional trade agreements that have been expanding rapidly in recent years. Because the twelve APEC member states involved in TPP and the EU countries of TTIP are chief export markets and foreign-trade sources of China, the TPP and TTIP led by the US have brought great pressure on China in the realm of international economic and trade cooperation.

To ensure China plays its active role in regional cooperation in the world, the Chinese leaders have proposed the Silk Road Economic Belt and the 21st Century Maritime Silk Road Initiative. The One Belt and One Road strategy, by enhancing connectivity among Asia, Europe, Africa and neighboring seas, and by establishing and consolidating the inter-connected partnership of countries along the One Belt and One Road, is designed to promote the diverse, independent, balanced and sustainable development in these regions.

4.1.2.2 Alleviating China's Problems in Energy Security and Imbalanced Regional Economic Development

China is a large importer of crude oil, which mainly sourced from Middle East countries, and Sudan in Africa; 80 % of crude oil is imported to China via the Malacca Strait. In recent years, because of the tension in South China Sea, security of crude oil import into China has been confronted with serious challenges. Along the One Belt and One Road, there are many major oil-producing countries and important oil-transport passages such as Strait of Hormuz and Strait of Malacca; hence, the Initiative can effectively alleviate the bottleneck problem faced by the crude oil import to China.

China proposed the One Belt and One Road strategy for another notable cause, i.e., to narrow down the gap between Central China, Western China and Eastern China in economic development. The Central and Western China accounts for 80 % of China's territory and has nearly 60 % of the nation's population. However, the import and export volume in this region only amounts to 14 % of the national total volume, its FDI accounts for 17 % of the national total, its external investment is at

---

[2] Doha Round refers to the round of multilateral trade negotiations among WTO members. In November 2001, the 4th WTO Ministerial Meeting held in Doha, the capital city of Qatar, kicked off the new round of multilateral trade negotiations (Doha Round). After Doha Round came to the end, the regional trade agreements have embraced prosperous development.

22 % of the total, and its GDP reaches only one third of the nation's total (Chinadaily 2014). According to the China National Statistics Bureau, in 2014, the per capita income of residents in the Eastern coastal provinces was 1.5–2 times higher than that in the Central and Western China. As a result, speeding up the economic growth in these lagging regions has been a major strategic goal of economic and social development in China.

4.1.3 Vision and Measures of the One Belt and One Road Initiative

On March 28, 2015, the National Development and Reform Commission, the Ministry of Foreign Affairs and the Ministry of Commerce jointly released the document *Vision and Actions on Jointly Building the Silk Road Economic Belt and 21st Century Maritime Silk Road,* which specified the vision, orientation, and actions of the One Belt and One Road strategy.

4.1.3.1 Vision of One Belt and One Road

According to the *Vision and Actions,* the One Belt and One Road strategy will strive to attain the following goals. (1) Realize better regional infrastructures and form safe and efficient network of land, maritime and air transport so that the regional connectivity is lifted to a new level; (2) Improve the convenience of investment and trade and establish better FTZ network so that the economic relations become closer and the political trust goes deeper in the region; and (3) Engage in more extensive cultural exchanges so that different civilizations learn from one another and peoples of all nations live in peace and harmony.

4.1.3.2 Measures for Development

According to the *Vision and Actions*, the One Belt and One Road strategy will focus on the cooperation in five aspects, namely communications of policies, connectivity of infrastructures, smooth relations of trade, freer flow of capitals and mutual understanding among peoples.

As per policy communications, the strategy will enhance inter-governmental cooperation among the countries along the One Belt and One Road, work actively to build the multi-level inter-governmental macro-policy communication and exchange mechanism, deepen the integration of interests, promote mutual political trust, and reach for the new consensus on cooperation.

As per infrastructure connectivity, the strategy, while respecting the sovereignty and security concerns of relevant countries, will encourage the countries along the

One Belt and One Road to integrate their infrastructure plans and technical standard systems, strive to push forward development of international trunk passages and gradually form the infrastructure network that connects sub-regions in Asia, Europe and Africa.

In terms of smooth trade relationship, the strategy will work to solve the problems in investment and trade facilitation, eliminate the investment and trade barriers, establish the positive pro-business environment within the region and the countries, actively consult with countries and regions along the One Belt and One Road about establishing FTZ, stimulate and release the potential cooperative strength. At the same time, it will endeavor to expand the scope of international trade, optimize the trade structure, explore new growth points of trade and promote trade balance.

In terms of capital flows, the strategy will deepen financial cooperation, push forward the development of Asian currency stability system, investment and financing system, and credit system, expand the scale and scope of bilateral currency swap and settlement among countries along the One Belt and One Road, and promote the openness and development of bond market in Asia.

In terms of the people's understanding, the strategy will preserve and carry forward the spirit of friendship and cooperation along the Silk Road, while engaging in extensive cultural and academic exchanges, talents exchange and cooperation, media cooperation, youth and women exchange, and volunteer service so as to build up the solid foundation of public consensus for bilateral and multilateral cooperation.

4.2 Economic Development Along the One Belt and One Road Region

Most of the countries along the One Belt and One Road are emerging economies and developing countries, which have recorded fast economic growth in recent years. Because these countries and China are complementary in resources and industrial structure, their trade relations with China have been on the rise continually, holding huge potential for cooperation.

4.2.1 Economic Development of Major Regions Along the Silk Road Economic Belt

The five Central Asian Countries constitute the major region along the Central Route and the Northern Route of the One Belt. The Northern Route also includes Mongolia and Russia; the Southern Route of the One Belt mainly runs through Myanmar, India, and Pakistan.

4.2.1.1 Major Economic Data of Countries Along the One Belt

Table 4.1 show the economic data of main countries along the One Belt in 2013.

The Five Central Asian Countries constitute the major region in the One Belt development. In recent years, GDP in the Five Central Asian Countries has been growing evidently faster than that of the international average. The GDP growth momentum and growth rate in the Five Central Asian Countries for 2004–2013, are shown in Fig. 4.2.

Table 4.1 Economic data of main countries along the One Belt (Year 2013)

	Country	GDP(million USD)	GDP per capita (USD)		Country	GDP(million USD)	GDP per capita (USD)
The five Central Asian Countries	Kazakhstan	231,876	13,608	The main countries in the Southern route of One Belt	Myanmar	N/A	N/A
	Kyrgyz	7226	1263		India	1,875,141	1499
	Tajikistan	8508	1037		Islamic Republic of Pakistan	232,287	1276
	Turkmenistan	41,851	7987	The main countries in the Northern route of One Belt	Mongolia	11,516	4057
	Uzbekistan	56,796	1878		Russia	2,096,777	14,612

Source: Compiled from data by the World Bank http://data.worldbank.org.cn/country/

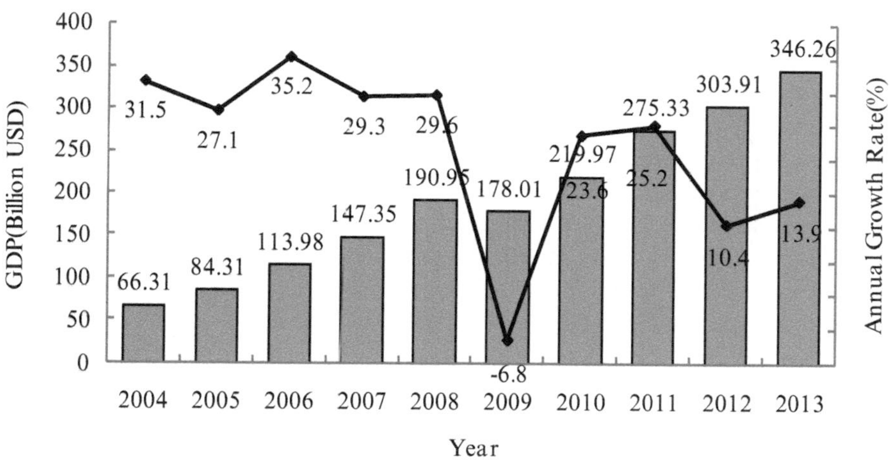

Fig. 4.2 GDP and growth rate of the Five Central Asian Countries for 2004–2013
Source: Compiled from data by the National Bureau of Statistics of China http://data.stats.gov.cn/workspace/index?m=gjnd

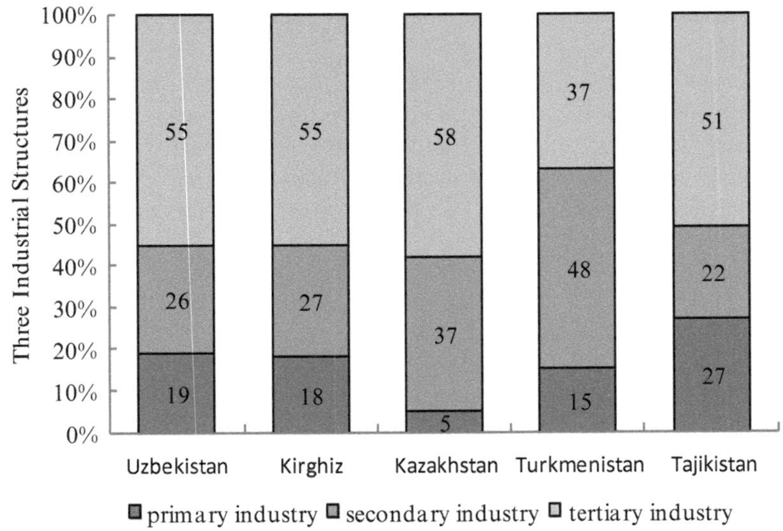

Fig. 4.3 The Five Central Asian Countries' three industrial structures for 2013
Note: Because the industrial data of Turkmenistan are only updated to 2012 by the WB, data shown here for Turkmenistan are for 2012
Source: Compiled from data by the World Bank. http://data.worldbank.org.cn/indicator/

4.2.1.2 Development of Major Industries in the Countries Along the One Belt

The Five Central Asian Countries have abundant mineral resources, though their processing industries lag behind. After the former Soviet Union broke up, the Five Central Asian Countries have been striving to end the economic structure dominated solely by the energy industries, gradually optimize their industrial structures, and raise the proportion of the tertiary industry. Apart from Turkmenistan, the tertiary industry in all other four countries has attained a level higher than 50 % of its respective total GDP. Figure 4.3 shows the proportions of three industries in the Five Central Asian Countries, and Fig. 4.4 shows the trend of the proportion change.

Figure 4.5 shows the proportions of the three industries in major countries (India and Pakistan) in the Southern Route of the One Belt and those (Mongolia and Russia) in the Northern Route. Similar to the Five Central Asian Countries, the proportion of tertiary industry in each of these countries stands higher than 50 %, while the proportion of primary industry is relatively low.

4.2.1.3 Trade Relations Between China and Countries Along the One Belt

In recent years, the year-on-year growth of trade volumes between China and the countries along the One Belt has been on the rise. Table 4.2 illustrates the trade volume between China and the countries along the One Belt for 2013. Of the

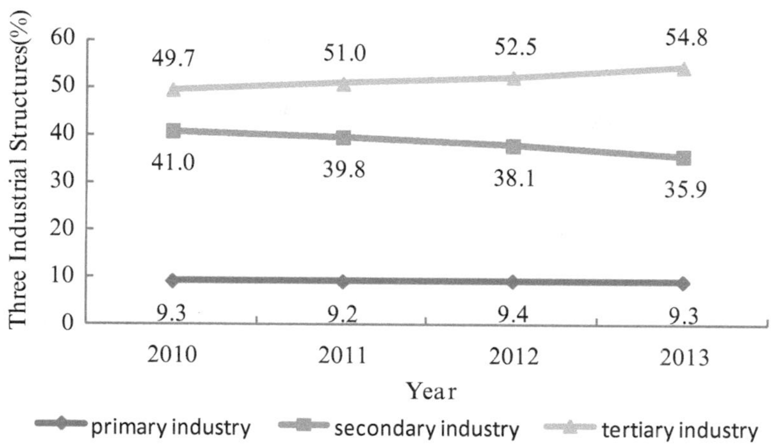

Fig. 4.4 Trends of weighted average proportions of the three industries for the Five Central Asian Countries
Note: Computed by using the arithmetic mean method based on the proportion of individual country's GDP in the total GDP of the Five Central Asian Countries
Source: Compiled from World Bank http://data.worldbank.org.cn/indicator/

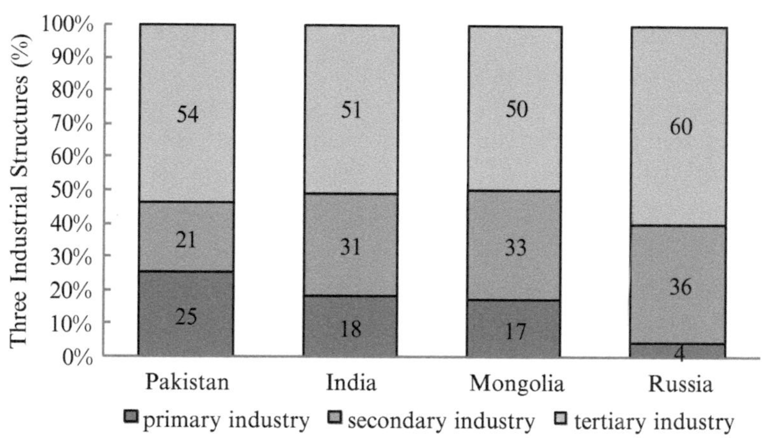

Fig. 4.5 Proportions of three industries in major countries in the Southern Route and the Northern Route of One Belt for 2013
Source: Compiled from data by the World Bank http://data.worldbank.org.cn/indicator/

Five Central Asian Countries, Kazakhstan has the largest trade volume with China, accounting for 57 % of the total trade volume between China and these countries. Turkmenistan and China have been enhancing their trade relations and the three-year average growth rate of trade volume between the two countries reached 180 %. Besides, China also has very large trade volumes with India, Pakistan and Russia.

Table 4.2 Trade volume and growth rate between main countries in "One Belt" and China for 2013

	Countries	Trade volume (100 million USD)	Average growth rate of trade volume in three years (%)
The Five Central Asian Countries	Kazakhstan	286.0	13
	Kyrgyz	51.4	7
	Tajikistan	19.6	12
	Turkmenistan	100.3	180
	Uzbekistan	45.5	28
The main countries in the South Route of One Belt	India	654.0	2
	Myanmar	102.0	43
	Islamic Republic of Pakistan	142.2	21
The main countries in the North Route of One Belt	Mongolia	59.6	16
	Russia	892.6	20

Source: Compiled from data by the National Bureau of Statistics of China http://data.stats.gov.cn/easyquery.htm?cn=C01

4.2.1.4 Major Categories of Goods Traded Between China and Countries Along the One Belt

Categories of goods traded between China and countries along the One Belt are evidently complementary with the country's respective industrial structure. Table 4.3, presents an overview of the major industries and export and import products for the Five Central Asian Countries, Russia, Mongolia, India and Myanmar.

4.2.2 Economic Development in Major Regions Along the 21st Century Maritime Silk Road

4.2.2.1 Main Economic Indicators of the Regions Along the One Road

The One Road mainly involves the coastal economic belt, along which the countries have different national status, political system and natural resources. In general, these countries have relatively rapid economic growth in recent years. Table 4.4 lists the main economic indicators of these countries. Of these countries, the per capita GDP in Western Asian Countries is far higher than that of the other regions, and the annual growth rates of GDP of these countries are also higher than that in other regions. East Africa and North Africa have relatively weak economic foundation. However, their GDP growth rates have been quite high over the past few years.

Table 4.3 Major goods traded between main countries along the "One Belt" and China

	The five central Asian countries					The main countries in the South Rroute of One Belt		The main countries in the North Route of One Belt	
Countries	Kazakhstan	Kyrgyz	Turkmenistan	Tajikistan	Uzbekistan	India	Myanmar	Mongolia	Russia
Industrial status	Mainly raw material-based industries and agriculture and animal husbandry; processing industry and light industry are relatively backward	Processing of agricultural products is an important part of economy; rich in mineral resources, but lack oil	Mainly farming and animal husbandry; oil and gas are pillar industries	Nonferrous metallurgy is an important industry; abundant coal reserves; cotton industry is important in agriculture industry	Gold, cotton, oil and natural gas are pillar industries; agriculture, animal husbandry and mining well-developed; processing industry is lagging behind	Mainly agriculture; industrial base is weak	Agriculture is the foundation of the national economy; rich in forestry resources	Animal husbandry and mining are main sectors of economy	Raw material industry is the core of economy, light industry is relatively backward; aerospace, nuclear industry reach world advanced level
Main goods imported	Consumer goods, electronics products	Energy products, clothing, chemical products	Food, meat and light industrial products	Mainly chemical products, metals and related products	Energy products, food, chemicals, metals and related products	Machinery, communication equipment etc.	Set of machinery equipment, electronic products, textiles and chemical products	Mineral products, machinery, and food	Light industrial products, chemicals, metals and related products
Main goods exported	Energy products	Mainly precious metals, chemicals and agricultural products	Natural gas, petroleum products, leather, and cotton	Non-precious metals and related products	Energy products, light industrial products, metals and related products	Agriculture products	Logs, sawn timber, agricultural products and minerals	Mainly mineral products, textiles, and livestock products	Minerals, metals and related products, chemical products, and consumer goods

Source: Compiled from respective sources by this study

Table 4.4 Economic indicators of the main areas along the One Road for 2013

Area/economic indicators	Population (million)	GDP (100 million USD)	GDP per person (USD)	Average annual growth rate of GDP for 2010-and 2013
Ten countries of ASEAN	615.68	23,527.47	3821.38	6.69–16.11
Six countries in South Asia	1612.89	23,285.56	1443.71	2.50–10.67
Eight countries in West Asia	255.39	28,625.97	11,208.73	3.99–18.30
Seven countries in East Africa and North Africa	344.06	4910.79	1427.30	0.47–16.66
Total	2828.02	80,349.79	2841.20	

Source: Compiled from data by the World Bank
http://data.worldbank.org.cn/country

4.2.2.2 Industrial Development of Major Countries Along the One Road

The One Road involves many countries. South Asia is not only a major region along the One Road, but also a major region in the Southern Route of One Belt. South Asia is the important "filling station" for China to reach Africa via the Indian Ocean. It has great potential of cooperation with China in marine resources, energy cooperation and development of port infrastructures. Currently there are several regional economic cooperation and development strategies, such as the "Bangladesh-China-India-Myanmar Economic Corridor," and the "China-Pakistan Economic Corridor." Even more, Pakistan is the essential hub connecting One Belt to the One Road.

Figure 4.6 exhibits the industrial structure of the Southern Asian Countries, which also shows that the proportion of secondary industry in the national economy of most Southern Asian countries is relatively low.

Pakistan is China's most important partner in South Asia in implementing the One Belt and One Road strategy. Though having abundant natural resources, Pakistan has a relatively slow industrialization process, whose proportion of tertiary industry is always far higher than that of the primary and secondary industry. See Fig. 4.7 for the structure of three industries in Pakistan for 2010–2013.

4.2.2.3 Trade Between China and Major Countries Along the One Road

The trade between China and major countries along the One Road plays an important role in the foreign trade in China. In 2013, the trade volumes between China and the ten ASEAN member states, six Southern Asian Countries, eight Western Asian Countries, and seven Eastern African and Northern African Countries reached 443.60 billion USD, 93.69 billion USD, 245.37 billion USD and 25.67 billion USD, respectively, accounting for 9.04 %, 2.25 %, 5.89 % and 0.62 % of the total foreign trade volume in China, and amounting to 17.8 % in aggregate. See Table 4.5 for details.

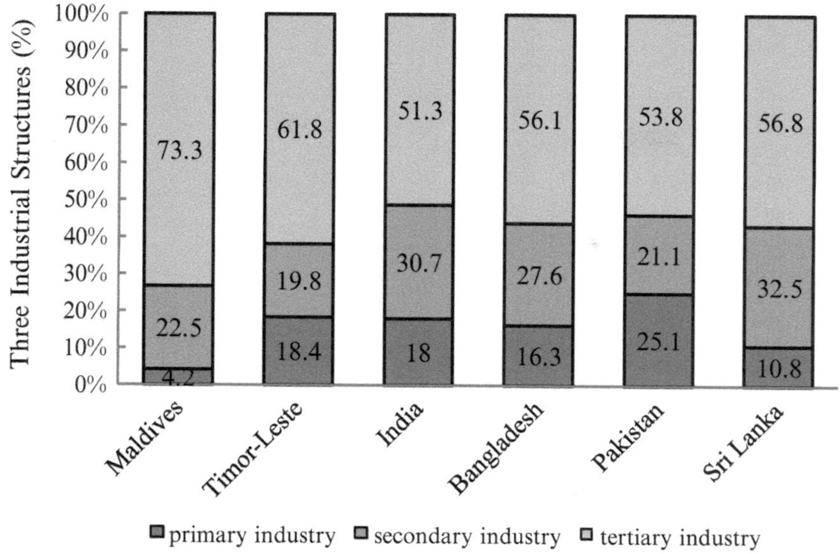

Fig. 4.6 The industrial structure of six South Asian countries for 2013
Note: Because the industrial data of Maldives and East Timor are only updated to 2012 by the WB, the data of these two countries shown here are those for 2012
Source: Compiled from data by the World Bank http://data.worldbank.org.cn/indicator/

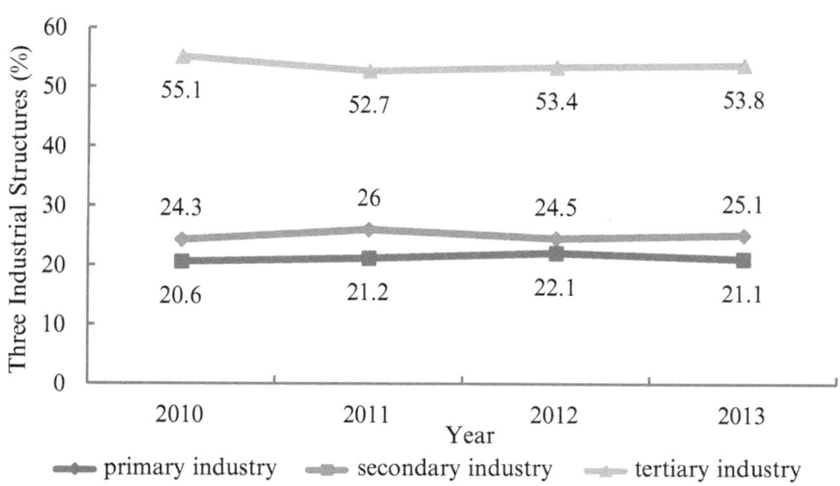

Fig. 4.7 Pakistan's three industrial structures for 2010–2013
Source: Compiled from data by the World Bank http://data.worldbank.org.cn/indicator/

As shown in Table 4.5, ASEAN is the region that has largest trade volume with China along the One Road. Figure 4.8 shows the import and export between ASEAN and China for 2006–2013. In 2006, the trade volume between ASEAN and China was only 150.9 billion USD, but that figure shot up to 443.6 billion USD in 2013.

Table 4.5 Trade status between areas along the One Road and China for year 2010–2013 (Unit: Billion USD)

Areas\year	2010	2011	2012	2013
Ten countries of ASEAN	292.86	363.09	400.15	443.60
Six countries in South Asia	79.69	96.04	90.64	93.69
Eight countries in West Asia	146.53	208.96	223.77	245.37
Seven countries in East Africa and North Africa	21.33	27.14	21.87	25.67

Source: Compiled from data by the National Bureau of Statistics of China http://data.stats.gov.cn/workspace/index?m=hgnd

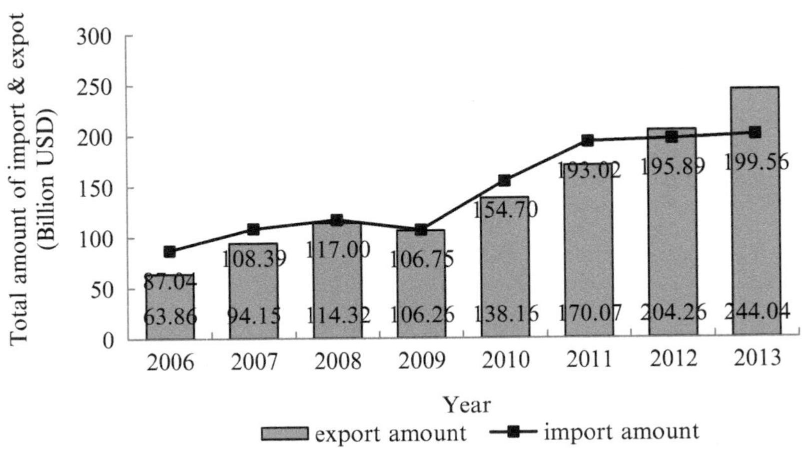

Fig. 4.8 Import-export trade on China-ASEAN for 2006–2013
Source: Compiled from data by the National Bureau of Statistics of China http://data.stats.gov.cn/workspace/index?m=hgnd

4.2.2.4 Major Categories of Traded Goods Between China and Countries Along the One Road

Major categories of goods in the trade between China and the countries along the One Road include oil products, chemical products, non-ferrous metals such as iron and steel, electronic products, and grains. Of these countries, ASEAN is China's largest trade partner along the One Road. As the bilateral trade relations develop rapidly, the trade structure between them is also changing. The proportion of primary products such as agricultural products is gradually declining, while that of high-tech products such as integrated circuits, mobiles phones and components, computers and components are increasing rapidly. See Figs. 4.9 and 4.10 for the import and export volume of electronic products and agricultural products between them.

China, as the world's second largest oil-importing country, has high dependence on imported oil and gas. In 2014, the following countries were largest exporters of crude oil to China, i.e., Saudi Arabia, Angola, Russia, Oman, Iraq and Iran; together they exported 68 % of China's total crude oil import, as shown in Table 4.6. Apart from Russia, the other five countries all lie along the One Road.

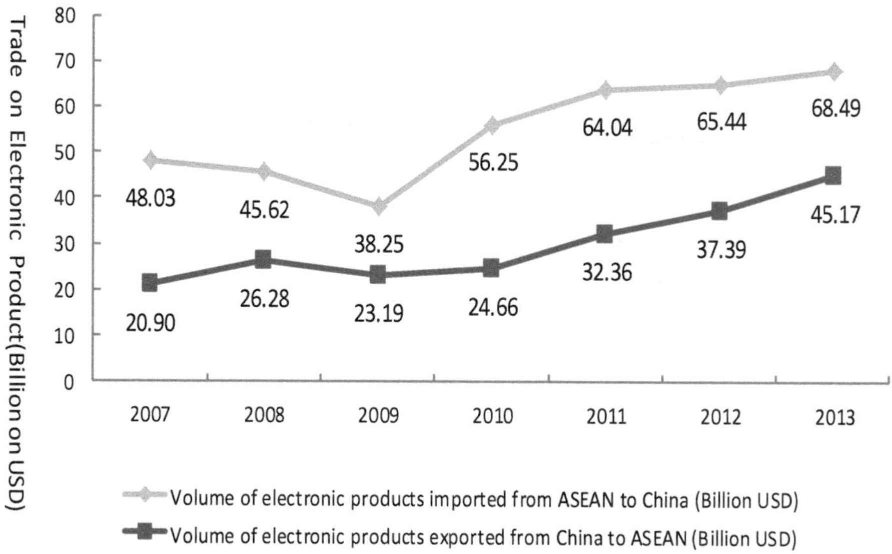

Fig. 4.9 Import-export trade of electronic products between China and the ASEAN for 2007–2013
Source: Compiled from the United Nations Commodity Trade Statistics Database http://comtrade.un.org/db/mr/daCommoditiesResults.aspx?px=H3andcc=85

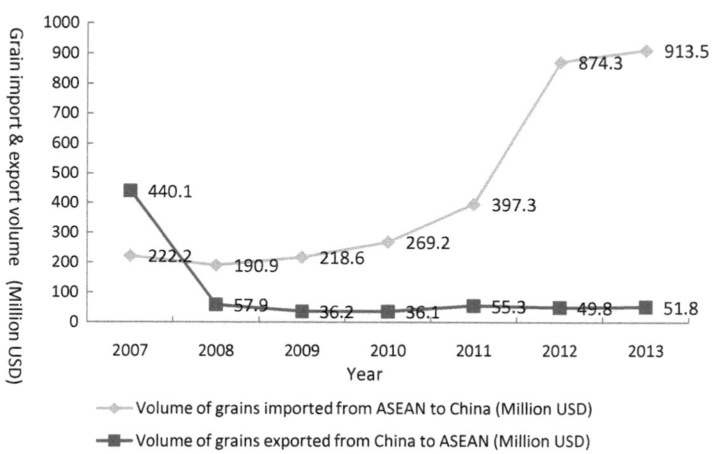

Fig. 4.10 Import-export trade of grains between China and the ASEAN for 2007–2013
Source: Compiled from the United Nations Commodity Trade Statistics Database http://comtrade.un.org/db/mr/daCommoditiesResults.aspx?px=H3andcc=10

Table 4.6 China's oil imports from some countries in year 2014

Country	Quantity (million tons)	Proportion (%) to total import
Saudi Arabia	49.67	16.11
Angola	40.65	13.18
Russia	33.11	10.74
Oman	29.74	9.65
Iraq	28.58	9.27
Iran	27.46	8.91
Total	209.21	67.86

Source: 《2014 China's import and export of crude oil inventory》 from the National Oil and Chemical Net http://www.cpcia.org.cn/news/hyfx/2015-1/144924.shtml. 2015-1-28

4.3 Development of Logistic Infrastructures Along the One Belt and One Road

Countries along the One Belt and One Road and China differ greatly in industrial structure and trade structure. Both parties have relatively strong economic complementarities. Consequently, as the One Belt and One Road strategy is implemented, there bound to be greater potential for the need of inter-regional logistic growth. Due to the geographical conditions, regional logistic service along the One Belt mainly adopts the land transport and that along the One Road mainly resorts to the maritime transport.

4.3.1 Development of Logistic Infrastructures Along the Silk Road Economic Belt

Trade between countries along the One Belt and China is mainly done via railway, highway and pipeline transport. Railways, highways and inland Customs ports are major components in the cross-border logistic systems and infrastructures along the One Belt.

4.3.1.1 Railway Network Development Along the One Belt

Railway transport is an important transport means between China and countries along the One Belt. An analysis of the 2012 data of railway network facilities in major countries along the One Belt reveals that these countries differ considerably in railway network density and the GDP value carried on unit railway mileage. In particular, Russia outperforms the other countries in terms of the GDP carried on unit railway mileage. See more details in Table 4.7.

Table 4.7 Transportation network conditions of main countries along the One Belt in year 2012

Country	Total railway mileage (km)	Railway mileage per unit area (km/1000 km^2)	Railway mileage per unit population (km/1000 persons)	GDP carried by unit railway mileage ($1000/km)
Kazakhstan	14,319	5.25	85.0	14,210
Kyrgyz	417	2.10	8.0	15,530
Tajikistan	621	4.34	8.0	12,290
Turkmenistan	3115	6.34	46.0	11,290
Uzbekistan	4192	9.37	15.0	12,190
India	64,460	19.60	5.0	28,840
Pakistan	7791	9.79	4.3	28,900
Russia	84,249	4.93	59.0	2,391,450
Mongolia	1818	1.16	64.0	5650

Source: Compiled from data by the World Bank
http://data.worldbank.org.cn/indicator/IS.RRS.TOTL.KM

Table 4.8 Transportation network conditions of China's main provinces along the One Belt in year 2013

	Province	Total railway mileage (km)	Railway mileage per unit area (km/1000 km^2)	Railway mileage per unit population (km/1000person)
Provinces alongside the Northern and Central routes of the Belt	Xinjiang	4,700	2.80	21
	Shaanxi	4,400	21.40	12
	Ningxia	1,300	19.60	20
	Qinghai	1,900	2.60	33
	Gansu	2,600	6.10	10
Provinces alongside the Southern route of the Belt	Guangxi	4,000	16.80	8
	Yunnan	2,600	6.60	6
Russia and Mongolia Economic Circle	Heilongjiang	6,000	12.70	16
	Jilin	4,400	23.50	16
	Liaoning	5,100	34.50	12
	Inner Mongolia	10,200	8.60	41
	Beijing	1,300	79.30	6

Source: Compiled based on data from the National Bureau of Statistic http://www.stats.gov.cn/

In general, the railway density in countries along the One Belt was lower than that in the provinces and autonomous regions of Mainland China along the One Belt in 2013. See more details in Table 4.8.

4.3.1.2 Major Inland Customs Ports in China and Countries Along the One Belt

There are many border ports in China that have a long history. And these border ports, since ancient times, have been the forefront for China to engage in economic and trade cooperation with countries along the Belt. Of these ports, the Manzhouli Port, Erenhot Port, Alashankou Port, Horgos Port and Ruili Port have been developing rapidly with large throughput of goods. Table 4.9 shows the development of these ports.

4.3.2 Development of Logistic Infrastructures Along the 21st Century Maritime Silk Road

Ports are main nodes connecting countries along the One Road and the navigation routes are important passages providing the connectivity of the One Road. Presently, China has established many cooperation projects on ports and navigation routes with countries along the One Road to facilitate transportation and trading.

4.3.2.1 Major Overseas Ports Along the One Road

The One Road connects nearly 50 large or medium-sized ports in China, East Asia, South Asia, West Asia and Africa. Table 4.10 shows a list of major overseas ports. Since maritime transport is the principal content of cooperation in the One Road initiative, port construction is the key point in its planning and implementation.

To better realize connectivity of infrastructures between China and countries along the One Road, China has helped to build ports in the key passages or at the key nodes. Table 4.11 shows the ports China invested in or helped build in recent years.

4.3.2.2 Construction of Coastal Ports in China

Coastal ports in China afford the key strategic support for the One Road initiative. The *Vision and Actions on Jointly Building Silk Road Economic Belt and 21st-Century Maritime Silk Road* proposed "enhancing construction of ports in coastal cities including Shanghai, Tianjin, Ningbo, Zhoushan, Guangzhou, Shenzhen, Zhanjiang, Shantou, Qingdao, Yantai, Dalian, Fuzhou, Xiamen, Quanzhou, Haikou and Sanya." These ports have different positioning in the planning and execution of the One Road Initiative, as shown in Table 4.12. Fujian Province is the core region in the One Road and Xiamen Port, as the largest port in Fujian Province, will play an essential role in the implementation of the strategy.

Table 4.9 Main inland ports in China along the Silk Road Economic Belt

Inland port	Manzhouli port	Erenhot port	Alashankou port	Horgos port	Ruili port
Docking countries	Russia, Mongolia	Mongolia	Kazakhstan	Kazakhstan	Myanmar
Development status	Main business: Storage, processing, bonded, trading and other functions; clearance for containers, coal, automobiles, chemical products	Main business: China's only railway crossing leading to Mongolia; 70 % of Mongolia's fruits, vegetables and daily necessities pass through the port. It will develop into a free border economic cooperation zone	Main business: Railway crossings, road crossings, crude oil pipeline transportation; trade, processing, storage and other functions	Main business: Trade, processing, warehousing, and tourism	Main business: immigration center, clearance of goods, garages, border management as one of the modern port management functions
	Throughput volume: China's largest land border crossings; annual capacity could reach 70 million tons	Throughput volume: The city's import and export cargo in 2014 reached 8.54 million tons, valued at $ 2.24 billion USD	Throughput volume: The volume amounted to 25.45 million tons in 2014	Throughput volume: With the clearance capacity of 2 million tons of goods and 3 million passenger-times	Throughput volume: 3.465 million tons cleared in 2014
	Main goods: Timber, oil, steel and fruit account for more than 70 % of total import and export volume		Main goods: Metal ores, oil, containers and mineral materials, account for over 60 % of total import and export		

Source: Compiled from related sources by this study

Table 4.10 Main ports in areas alongside the One Road

Areas alongside the One Road	Main ports
Ten ASEAN countries	Singapore Port, Port Kelang, Jakarta Port, Manila Port, Da Nang, Ho Chi Minh Port, Tanjung Priok Port, Malacca Port, Sihanoukville Port
Six countries in South Asia	Colombo Port, Calcutta Port, Male Port, Gwadar Port
Eight countries in West Asia	Aden Port, Kuwait Port, Dubai Port, Dammam Port
Seven countries in East Africa and North Africa	Mombasa Port, Maputo Port, Dar es Salaam Port, Mogadishu Port, Nairobi Port, Port Said

Source: Compiled from various sources by this study

Table 4.11 Ports invested by China along The 21st Century Maritime Silk Road

Country	Port project	Chinese companies	Stake held by Chinese companies or other rights and interests
Singapore	Container Terminal	COSCO	49 %
Malaysia	Kuantan Port	Guangxi Beibu Gulf International Port Group, Ltd.	40 %
Sri Lanka	Colombo South Container Terminal	CMHI (China Merchants Holdings (International) Co., Ltd.)	85 %
	Hambantota Port Phase II container terminal	CMHI, CHEC (China Harbour Engineering Company, Ltd.)	64.98 %
Myanmar	Madrid Kyaukpyu Port Island	CNPC	50.90 %
Pakistan	Gwadar	China Overseas Port Holdings Limited	40-year concession
Egypt	Port Said	COSCO	20 %
	Aiyinsuhe Port	CHEC	Unknown
	Damietta Port	CHEC	Unknown
Israel	Haifa Newport Pier	SIPG (Shanghai International Port (Group) Co., Ltd.)	25-year concession
Greece	Piraeus Pier 2,3	COSCO	35-year concession
	Piraeus Pier 1	COSCO	67 %(Yet to be finalized)
Italy	Naples Harbour Container Terminal	COSCO	50 %
Belgium	Zeebrugge container terminal	SIPG	25 %
	Antwerp Container Terminal	COSCO	25 %

Source: "Ports become key nodes in the 21st Century Maritime Silk Road and Chinese Investment Layout along Relative Countries," 21st Century Business Herald, March 31, 2015

Table 4.12 Positioning of major ports in China's coastal areas in One Road plan

	Ports	Position
Bohai port cluster	Tianjin Port	Lie in the junction of Mongolia and Russia economic corridors and the New Eurasian Continental Bridge Economic Belt; rendering support to the Beijing-Tianjin-Hebei integration strategy
	Qingdao Port	
	Dalian Port	
	Yantai Port	
Yangtze River Delta port cluster	Shanghai Port	Located in the ocean access of the Yangtze River Economic Belt and the New Eurasian Continental Bridge Economic Belt
	Ningbo Zhoushan Port	
Pearl River Delta port cluster	Guangzhou Port	Adjacent to the Strait of Malacca and linking with the Indochina Economic Corridor
	Shenzhen Port	
	Xiamen Port	
	Fuzhou Port	
	Quanzhou Port	
	Shantou Port	
	Zhanjiang Port	
	Haikou Port	
	Sanya Port	

Source: "Three port clusters share benefit from 'One Belt and One Road'," from Cnship Net http://www.chinaship.cn/policy/2015/0419/3663.html. 19 Apr. 2015

4.3.2.3 Development of Navigation Routes

The One Road not only includes the near-sea, but also ocean routes. The passages and navigation routes of trade between China and SE Asian and Southern Asian countries are relatively short and development have been rapid in recent years due to the frequent trade exchanges. Of the ocean routes, the Persian Gulf route is the most important oil transport passage for China. The navigation route runs through the Persian Gulf, Strait of Hormuz, Indian Ocean, Strait of Malacca, South China Sea and the Pacific before reaching China. The Strait of Hormuz and the Strait of Malacca along the Persian Gulf route are important nodes along the route. See Table 4.13 for main routes in the One Road.

4.4 An Outlook on the Regional Logistic Development in the One Belt and One Road

The strategy of One Belt and One Road will bring vast opportunities for the development of logistic industry in the covered regions. Within the scope of One Belt and One Road, which connects Europe and Asia, and joins the countries along the West Pacific and Indian Ocean, a cross-border logistic system will be formed. In the future, connecting Western China to the countries along the One Belt via highways and railways, and port construction in countries along the One Road in SE Asia and

Table 4.13 Main routes in the 21st Century Maritime Silk Road

	Routes	Main ports of call
Near ocean routes	Hongkong-Macao	Hongkong, Macao
	Singapore-Malacca port	Singapore port, Port Klang, Penang, Malacca port
	Siam Bay route	Hai Phong port, Sihanoukville Port, Bangkok Port
	Colombo route, Bay of Bengal line	Colombo Port, Yangon port; Chittagong, Calcutta
	Indonesia line	Jakarta port, port of Semarang
	Persian Gulf line	Karachi port, Bandar Abbas, Basra port, Kuwaitport, Dammam
Ocean route	China to Red Sea route	Aden, Chittagong, Aqaba port, Port Sudan
	China to East Africa route	Mogadishu port, the port of Mombasa, Dar es Salaam, Maputo harbor, etc.
	China to the Mediterranean route	Istanbul Port, Venice, Genoa Port, Marseille, Port of Barcelona, Valencia Port, Alexandria, Tripoli, Benghazi Port, Tunisia Port, Algiers Port etc.

Source: Compiled from various sources by this study

South Asia, will be the focus of logistic infrastructure development along the One Belt and One Road. The FTZ development and investment and trade facilitation measures in China will also help improve the operating efficiency of logistic system in the One Belt and One Road.

4.4.1 Transport Infrastructures Are Focus of Regional Logistic System in the One Belt and One Road

Connectivity is the crucial foundation for achieving the grand vision of One Belt and One Road. At present, the transport infrastructures including highways and railways in Mongolia, the Five Central Asian Countries, SE Asian and Southern Asian Countries along the One Belt and One Road are still lagging behind. China will cooperate with these countries and undertake the transport infrastructure development as the steering project, so as to build the regional logistic system featuring connectivity, thus providing fundamental support for economic and trade cooperation. Consequently, highways and railways will become the focal development in regions along the One Belt, and ports will be the focal development in regions along the 21st Century Maritime Silk Road.

The provinces and autonomous regions where the One Belt and One Road envelops have all drafted development plans of infrastructures to meet the needs of this strategy. For example, Urumqi in Xinjiang is accelerating its establishment of the Five Centers in the core area of the Silk Road Economic Belt.[3] Coastal ports in

[3] The Five Centers are regional transport hub center, commerce and logistics center, financial service center, cultural high-tech center, and medical service center.

China are also actively developing the maritime-railway joint transport to build the logistic passages to and from sea.

Internationally, China has begun to invest in and build highways, railways, pipelines, border ports, seaports, navigation routes, bridges and other infrastructures in Central Asia, South Asia and SE Asia among others. For instance, the D Line of China-Central Asia Natural Gas Pipeline Project, which is the first major strategic project invested by China in Central Asia, was initiated in Dushanbe, the capital city of Tajikistan, in September 2014. In April 2015, the Chinese President Xi Jinping signed 51 agreements of cooperation when he made the state visit to Pakistan.

4.4.2 Operations Efficiency of Regional Logistic System in the One Belt and One Road Region Will Be Effectively Improved

The regional logistic system along the One Belt and One Road involves many countries which necessitates various transport means. Because these countries have different technical standards, management regulations, customs and habits, the operating efficiency in cross-border logistic system, to a large extent, relies on the coordination and communication efficiency among the sub-systems. The coordinated development of One Belt and One Road will vigorously push forward the development of networked regional logistic system.

At present, with the major cities along the One Belt and One Road as the nodes, and the highway, railway network and navigation routes as the connecting arcs, China is pursuing the coordination of regional logistic system, by facilitating construction of infrastructure hardware and improving the soft environment of management regulations. In other countries, China has dispersed 10 billion USD to establish the Maritime Cooperation Fund, which will be used to build infrastructures such as ports, railways and highways in countries along the One Belt and One Road. Furthermore, the Fund will be used to develop their e-commerce and improve the Customs clearance capability. Meanwhile, China will seek cooperation and coordination among the Customs administration, quality inspect and e-commerce departments to realize the coordinated standards of infrastructures and facilitation of clearance regulations.

Facilitation of Customs clearance and trade is the crucial factor which influences the operating efficiency of a logistic system. The FTZ and establishment of the "Single Window" practice in international trade will also effectively improve the Customs clearance efficiency in the One Belt and One Road. For instance, Shanghai, as the prominent node along the One Belt and One Road, is also the pilot test city of FTZ and the "Single Window" practice of international trade in China. Shanghai's Customs Office, together with the quarantine and inspection departments, has made pilot test operations of the "Three-Once" Program (i.e., "one application, one inspection and one clearance") in the Shanghai Customs Port. This program has laid a foundation for the establishment of the "Single Window" practice in countries along the One Belt and One Road. And the "Single Window" of international trade will become an important content in constructing the soft environment for the One Belt and One Road.

4.4.3 The One Belt and One Road Regional Logistic System Will Have Its Operating Layout Continually Optimized

The One Belt and One Road involves many countries with complex topography and various transport means. Due to the international relations and technical development level, some traditional logistic nodes and passages still have ample space for betterment.

For instance, the Gwadar Port in Pakistan controls the several key navigation routes originating from Africa and Europe, passing through the Red Sea, the Hormuz Strait and Persian Gulf to reach East Asia and the Pacific. It is the vital point connecting China to the Arabic Sea and the Hormuz Strait, as shown in Fig. 4.11. The Gwadar Port is endowed with the natural advantage of building a deep-water port and can be developed into the maritime transit station integrating transit, warehousing and transport. The China-Pakistan Railway and Highway, upon completion, will greatly shorten the transport mileage of shipping oil from the

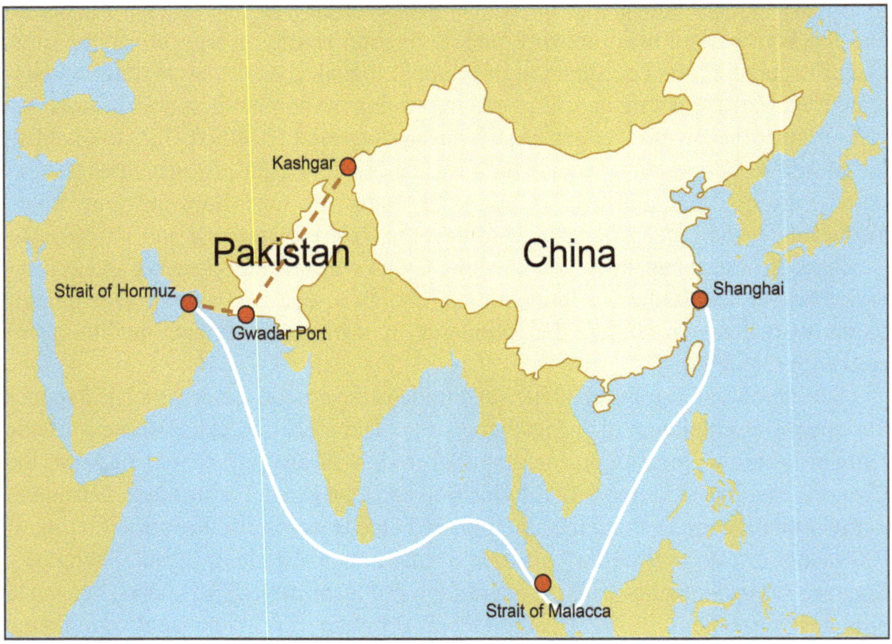

Fig. 4.11 The role of Gwadar in Pakistan Economic Corridor
Source: Compiled by this study

Middle East to China. To advance the development of Pakistan Economic Corridor, China and Pakistan have signed a number of infrastructure projects. The Gwadar Port is now being built as a major investment and construction project.

4.5 Summary

The One Belt and One Road Initiative is a major diplomatic and economic development strategy of China. The strategy will yield great impact on international trade and logistic activities in countries along the One Belt and One Road. Most of the countries along the One Belt and One Road regions are emerging economies or developing countries; they complement well with China in terms of natural resources, industrial structure and categories of goods in trading, and have huge potential of engaging in cooperation with China. In view of logistics, because of the geographical conditions, regional logistic service along the One Belt will mainly adopt the form of land transport, whereas the regional logistic service along the One Road will rely mainly on maritime shipping. Therefore China intends to focus on the following logistic infrastructure development when it implements the One Belt and One Road strategy: connect China with countries along the One Belt via highways and railways; enhance the port infrastructure development in countries along the One Road in SE Asia and South Asia. In the meantime, the FTZ development in China will facilitate its global investment and trade. These measures, implemented in cross-border clearance, will effectively improve the operating efficiency of logistic system in the One Belt and One Road region. In the future, with the improvement in logistic infrastructures and the advancement of system operating efficiency, the One Belt and One Road regions will form a cross-border logistic system that runs through Europe and Asia on the land, and connects countries along the West Pacific and Indian Ocean on the sea, thus laying a solid foundation for the implementation of the One Belt and One Road strategy.

Reference

Chinadaily (2014) Xi Jinping's visit to SE Asia and South Asia: fully implement the One Belt and One Road initiative. ChinadailyNet. http://world.chinadaily.com.cn/2014-09/17/content_18615669.htm. 17 Sep 2014

Chapter 5
Logistics Development in Beijing-Tianjin-Hebei Region

Lan-bing Li

Beijing-Tianjin-Hebei Region, as one of the most robust regions in China with greatest potential, is the third growth pole after Yangtze River Delta and Pearl River Delta. In 2014, the policy of Beijing-Tianjin-Hebei Coordinated Development became a major national strategy, instilling new dynamics into the region's economic and social development. Owing to its considerable economic scale and well-developed open economic system, the Beijing-Tianjin-Hebei region has generated huge logistic demands which resulted in a most developed logistic industry. The Beijing-Tianjin-Hebei Coordinated Development Strategy will provide superb opportunities for logistic development in the region while having far-reaching influence on it.

This chapter mainly analyzes the current logistic development in the Beijing-Tianjin-Hebei region. Section 5.1 describes the background, meaning and functional positioning of the Beijing-Tianjin-Hebei Coordinated Development Strategy. Section 5.2 analyzes the economic development characteristics of the Beijing-Tianjin-Hebei region. Section 5.3, in a systematic way, discusses the current development of logistic market in the region. Section 5.4 summarizes the profound influence of the Strategy on logistics of this region.

L.-b. Li (✉)
Logistics Research Center, Nankai University, Tianjin, People's Republic of China
e-mail: lilanbing1978@163.com

L. Wang et al. (eds.), *Contemporary Logistics in China*, Current Chinese Economic Report Series, DOI 10.1007/978-981-10-1052-1_5

5.1 Background, Meaning and Functional Positioning of the Beijing-Tianjin-Hebei Coordinated Development

5.1.1 Background

The Beijing-Tianjin-Hebei Coordinated Development is currently a major national strategy of China, bearing upon the nation's economic development as a whole. It is not limited to the regional cooperation among Beijing, Tianjin and Hebei, but also carries on the significant task of promoting China's in-depth reform, transition and development, and of catching up with the nation's most developed regions.

The Beijing-Tianjin-Hebei Coordinated Development can be traced back to the preliminary stage of the reform and opening-up program in China. It has since evolved and taken shape over the past 30 years. In 1986, 15 cities in Circum-Bohai-Sea region established the Joint Mayors Meeting in Circum-Bohai-Sea Region. In 1988, Beijing and its six neighboring cities in Hebei Province (i.e., Baoding, Langfang, Tangshan, Qinhuangdao, Zhangjiakou and Chengde) co-established the "Circum-Beijing Economic Coordination Region." In 1996, the "Circum-Bohai-Sea Economic Area" was listed in the 9th Five-year Plan for the National Economic and Social Development in China. Beijing proposed the concept of "Capital City Economic Circle," which centered in Beijing and included seven prefecture-level cities in Hebei Province. In 2005, the State Council Executive Meeting passed the "Beijing Urban Master Plan for 2004–2020," which stipulated that "it is imperative to enhance coordinated development of Beijing-Tianjin-Hebei region in industrial development and ecological construction," and "to basically establish the 2-h transport circle centered in Beijing and Tianjin." In 2006, the 11th Five-year Plan for the National Economic and Social Development in China raised the subject of Beijing-Tianjin-Hebei City Cluster development. In 2011, the 12th Five-year Plan for the National Economic and Social Development in China (2011–2015) specified "building the Capital City Economic Circle." In 2013, the Central Government convened the 2013 Beijing-Tianjin-Hebei Capital City Development Summit, which addressed the Beijing-Tianjin-Hebei regional coordinated development. By 2014, the Beijing-Tianjin-Hebei Coordinated Development Plan had been upgraded from a regional development strategy to a major national one.

5.1.2 Meaning and Functional Positioning

The significant meaning of Beijing-Tianjin-Hebei Coordinated Development is manifested in four aspects. First, Beijing-Tianjin-Hebei Coordinated Development is necessary for building the new Capital City Economic Circle and driving the innovation mechanism for regional development. Second, it is important for exploring and improving the layout and format of city cluster, and for providing demonstration and prototype for optimizing regional development. Third, it is essential for identifying the effective path towards ecological development and for promoting balanced development of population, economy, resources and environment. Fourth, it is vital for achieving complementary advantages in Beijing-Tianjin-Hebei, promoting Circum-Bohai-Sea Economic Area, and boosting the development of China's Northern hinterland (Chinanet 2014).

In July 2015, the *Outline of Beijing-Tianjin-Hebei Coordinated Development Plan* was issued, which clearly defined the overall positioning of Beijing-Tianjin-Hebei region as a whole and that of Beijing, Tianjin and Hebei, respectively. The overall positioning of Beijing-Tianjin-Hebei region reads: A world-class city cluster cored in Beijing, a block which steers the coordinated and holistic development and reform of the entire region, a demonstration region of ecological restoration and environmental improvement, and a new engine of innovation-driven growth across China. Beijing is positioned as the national political center, cultural center, center of international exchange and center of R&D and innovation. Tianjin is positioned as an advanced national manufacturing and R&D base, a kernel region in North China for international shipping, a demonstration region of financial innovation and operations, and a leading region in reform and opening-up. Hebei is positioned as the major national base of modern trade, commerce and logistics, a pilot-test region of industrial transformation and upgrading, a demonstration region of new urbanization and rural-urban coordinated development, and a supporting area for ecological and environmental development in Beijing-Tianjin-Hebei region.

5.2 Economic Development Characteristics in Beijing-Tianjin-Hebei Region

5.2.1 Beijing and Tianjin Well-Developed; Large Gap in Regional Development

There exists a relatively large gap among Beijing, Tianjin and Hebei in terms of economic and social development. For instance, the per capita gross product in Hebei province is far lower than that in Beijing and Tianjin. In 2014, the per capita gross product in Beijing, Tianjin and Hebei was 99,995 RMB, 103,655 RMB and 39,846 RMB, respectively, indicating that Hebei's GDP per capita is less than 40 % of that in Beijing or Tianjin, as shown in Fig. 5.1. Another example is that the urbanization level in Beijing and Tianjin reached 85.96 % and 79.44 %, respectively, while that in Hebei Province stands at a much lower 43.11 %. Furthermore, Heibei's urbanization level is 6.57 percentage points lower than the average level in China.

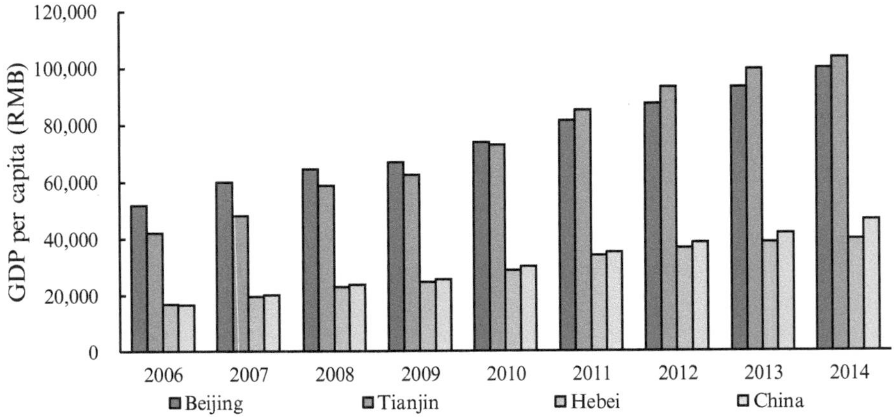

Fig. 5.1 GDP per capita of Beijing-Tianjin-Hebei region and China for 2006–2014
Source: Compiled from the *China Statistical Yearbook (2014)* and the *China Statistical Bulletin of National Economic and Social Development (2014)*, both published by the National Bureau of Statistics of China; the *Statistical Bulletin of National Economic and Social Development of Beijing, Tianjin and Hebei (2006–2014)*, published by the Bureau of Statistics and the General Research Team of National Bureau of Statistics of Beijing, Tianjin and Hebei

5.2.2 Different Stages of Industrial Development

Beijing, Tianjin and Hebei are in different stages of industrial development. In 2014, the proportions of the three industries in Beijing were 0.7:21.4:77.9; the proportion of tertiary industry was much higher than those of the primary and secondary industry, thus taking the dominant position. This means that Beijing's development is in the post-industrialization stage. The proportions of the three industries in Tianjin were 1.3:49.4:49.3; the secondary industry has the highest proportion, followed neck-and-neck by the tertiary industry. This means that Tianjin's development is in the later stage of industrialization. The proportions of the three industries in Hebei were 11.7:51.1:37.2; its secondary industry is in the dominant position and the proportion of tertiary industry is relatively low. This means that the development of Hebei Province is in the middle stage of industrialization. More details are shown in Table 5.1.

5.2.3 The Broken Urban System and Spatial State

Urban system development in Beijing-Tianjin-Hebei region is imbalanced; huge gap of development exists among cities in the region. In Beijing-Tianjin-Hebei region, there are 11 prefecture-level cities and two cities directly administered by the Central Government. As shown in Table 5.2, the region is composed of one

Table 5.1 Proportions of the three-tier industry of Beijing-Tianjin-Hebei region for 2006–2014

	Beijing			Tianjin			Hebei		
	Primary industry	Secondary industry	Tertiary industry	Primary industry	Secondary industry	Tertiary industry	Primary industry	Secondary industry	Tertiary industry
2006	1.3	27.8	70.9	2.7	57.3	40.0	13.8	52.3	33.9
2007	1.1	27.5	71.4	2.1	57.6	40.3	13.2	52.8	34.0
2008	1.0	23.6	75.4	1.9	60.1	38.0	12.7	54.3	33.0
2009	1.0	23.2	75.8	1.7	54.8	43.5	13.0	52.1	34.9
2010	0.9	24.1	75.0	1.6	53.1	45.3	12.7	53.0	34.3
2011	0.8	23.1	76.1	1.4	52.5	46.1	12.0	54.1	33.9
2012	0.8	22.8	76.4	1.3	51.7	47.0	12.0	52.7	35.3
2013	0.8	21.7	77.5	1.3	50.6	48.1	12.4	52.1	35.5
2014	0.7	21.4	77.9	1.3	49.4	49.3	11.7	51.1	37.2

Source: Compiled from the *Statistical Bulletin of National Economic and Social Development of Beijing, Tianjin and Hebei (2006–2014)*, published by the Bureau of Statistics and the General Research Team of National Bureau of Statistics of Beijing, Tianjin and Hebei

Table 5.2 City size of Beijing-Tianjin-Hebei region for 2014

City size (million people)	Number of cities		City name
	Value	Proportion (%)	
Super-Mega city (>10)	1	7.69	Beijing
Mega city (5–10)	1	7.69	Tianjin
Large city (1–5)	4	30.77	Baoding, Tangshan, Shijiazhuang, Handan
Medium-sized City (0.5–1)	6	46.15	Qinhuangdao, Zhangjiakou, Cangzhou, Xingtai, Chengde, Langfang
Small City (<0.5)	1	7.69	Hengshui

Source: Compiled from the *Statistical Bulletin of National Economic and Social Development of Beijing, Tianjin and all the prefecture-level cities in Hebei province (2014)*, published by the Bureau of Statistics and the General Research Team of National Bureau of Statistics of Beijing, Tianjin and all the prefecture-level cities in Hebei province

super-mega city[1] with more than 10 million population, one mega city with 5–10 million population, four large cities with 1–5 million population, six medium-sized population with 0.5–1 million population, and one small city with 0.5 million population. Beijing-Tianjin-Hebei region exhibits the pronounced features of a broken urban system, showing that the region is yet to establish a rational and orderly urban system. Except for Beijing and Tianjin, which are super-mega city and mega city respectively, the development of the many medium-sized and small cities in Hebei Province is lagging way behind.

5.2.4 Beijing and Tianjin Having Developed Export-Oriented Economy; Tianjin Free Trade Zone Began Operations

Both Beijing and Tianjin are cities with most developed export-oriented economy in China. In 2014, the total volume of foreign trade in Beijing, Tianjin and Hebei amounted to 415.7 billion USD, 133.9 billion USD, and 59.2 billion USD, respectively, as shown in Fig. 5.2. Their extent of foreign trade dependency was 121.14 %, 52.95 % and 12.65 %, respectively. In April 2015, the Tianjin Pilot Free Trade Zone went into operation, which would serve as the new carrier for Beijing-Tianjin-Hebei Coordinated Development and instill new vigor into the region's development of export-oriented economy.

[1]According to the *Notification of Adjusting Criteria of Categorizing Cities* issued by the State Council in 2014, cities with 0.5 million permanent urban population (PUP) and below are small cities, those with PUP ranging between 0.5 million and 1 million are medium-sized cities, those with PUP ranging between 1 million and 5 million are big cities, those with PUP ranging between 5 million and 10 million are mega cities, and those with PUP more than 10 million are super-mega cities.

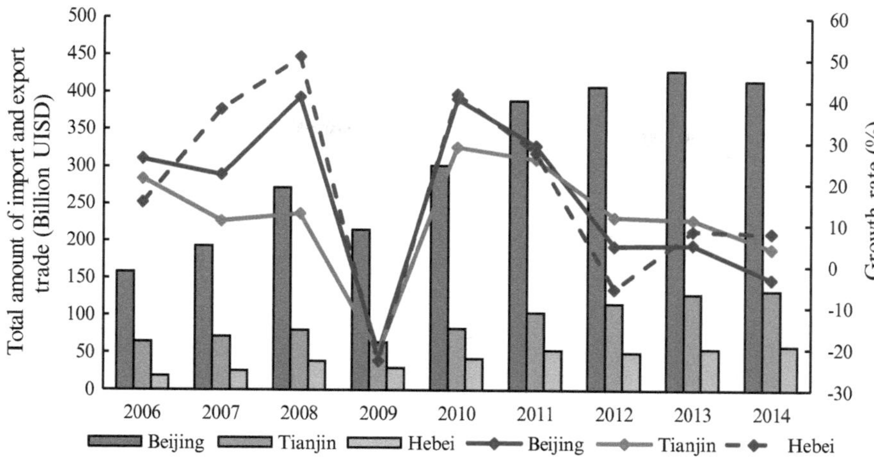

Fig. 5.2 Total import and export value and growth rate of Beijing-Tianjin-Hebei region for 2006–2014
Source: Compiled from the *Statistical Bulletin of National Economic and Social Development of Beijing, Tianjin and Hebei (2006–2014)*, published by the Bureau of Statistics and the General Research Team of National Bureau of Statistics of Beijing, Tianjin and Hebei

Table 5.3 Import dependence and export dependence of Beijing-Tianjin-Hebei region for 2006–2014

		2006	2007	2008	2009	2010	2011	2012	2013	2014
Import dependence (%)	Beijing	117.9	111.3	133.8	93.5	118.1	131.4	123.0	116.2	103.0
	Tianjin	55.3	48.3	39.6	30.8	32.7	33.6	33.0	34.3	32.2
	Hebei	4.0	4.8	6.3	5.5	6.5	6.6	5.0	5.2	5.1
Export dependence (%)	Beijing	37.3	37.8	35.9	27.2	26.6	23.4	21.1	20.0	18.2
	Tianjin	59.8	55.1	43.5	27.1	27.5	25.4	23.7	21.1	20.8
	Hebei	8.9	9.5	10.4	6.2	7.5	7.5	7.0	6.8	7.5

Source: Compiled from the *Statistical Bulletin of National Economic and Social Development of Beijing, Tianjin and Hebei (2006–2014)*, published by the Bureau of Statistics and the General Research Team of National Bureau of Statistics of Beijing, Tianjin and Hebei

In 2014, the extent of import dependency of Beijing, Tianjin and Hebei was 103.0 %, 32.2 % and 5.1 %, and the extent of export dependence was 18.2 %, 20.8 % and 7.5 %, respectively, as shown in Table 5.3. This shows that Beijing has a high extent of import dependency and low extent of export dependency; the latter has been decreasing year by year, telling that Beijing's economic growth relies heavily on import and its export trade has relatively weak role in stimulating the nation's economy. Apparently Beijing falls short of fulfilling the national strategy of developing export-oriented economy. In comparison, Tianjin has a more balanced extent of import and export dependency. Yet its extent of import dependency is still higher than that of the export dependency, with the latter extent diminishing. Hence Tianjin needs to strengthen its export performance to fulfill its positioning as the Bridge

Head in the One Belt & One Road strategy. For Hebei Province, the extent of export dependency is higher than that of the import dependency. However, both levels are relatively low, revealing the fact that Hebei Province is severely deficient in developing export-oriented economy.

5.2.5 *Beijing-Tianjin-Hebei Region Lagging Behind Yangtze River Delta and Pearl River Delta*

The Beijing-Tianjin-Hebei region is expected to become the third economic growth pole in China, following the Yangtze River Delta and Pearl River Delta.[2] Nevertheless, compared with these two leading regions, the Beijing-Tianjin-Hebei region lags behind them markedly in terms of economic development, as shown in Table 5.4. First, consider the per capita gross product. In 2014, per capita GDP in Pearl River Delta and Yangtze River Delta was 100,290 RMB and 81,040 RMB, respectively, while that in Beijing-Tianjin-Hebei was 60,150 RMB. Second, compare the total volume of foreign trade. In 2014, total volume of foreign trade in Beijing-Tianjin-Hebei region was 609.44 billion USD, and that in Yangtze River Delta and Pearl River Delta was 1782.36 billion USD and 1029.31 billion USD, respectively. Third, contrast the urbanization process. In 2014, urbanization rate in Yangtze River Delta and Pearl River Delta was 67.40% and 84.00%, respectively, but that in Beijing-Tianjin-Hebei region was only 61.10%.

Table 5.4 Comparison of economic development among Beijing-Tianjin-Hebei region, Yangtze River Delta and Pearl River Delta for 2014

	Beijing-Tianjin-Hebei region	Yangtze river delta	Pearl river delta
GDP (Billion RMB)	6647.45	12,880.32	5780.22
Population (Million people)	110.52	158.94	57.63
GDP per capita (Thousand RMB)	60.15	81.04	100.29
Total import and export value (Billion USD)	609.44	1782.36	1029.31
Foreign trade dependency (%)	56.29	84.96	109.34
Urbanization rate (%)	61.10	67.40	84.00

Source: Compiled from the *Statistical Bulletin of National Economic and Social Development of Beijing, Tianjin, Hebei, Shanghai, Jiangsu, Zhejiang, Guangzhou, Shenzhen, Zhuhai, Foshan, Huizhou, Dongguan, Zhongshan, Jiangmen and Zhaoqing (2014)*, published by the Bureau of Statistics and the General Research Team of National Bureau of Statistics of Beijing, Tianjin, Hebei, Shanghai, Jiangsu, Zhejiang, Guangzhou, Shenzhen, Zhuhai, Foshan, Huizhou, Dongguan, Zhongshan, Jiangmen and Zhaoqing

[2] According to *Statistical Bulletin of National Economic and Social Development of Guangdong Province (2014)*, the Pearl River Delta includes Guangzhou, Shenzhen, Zhuhai, Foshan, Huizhou, Dongguan, Zhongshan, Jiangmen and Zhaoqing.

5.3 Status of Logistic Market in Beijing-Tianjin-Hebei Region

5.3.1 Scale of Logistic Market

5.3.1.1 Added Value of Transportation, Warehousing and Postal Service Industry

During 2006–2014, the transportation, warehousing and postal service industry in the Beijing-Tianjin-Hebei region grew rapidly, as shown in Table 5.5. The average growth rate of the added value of the region was 12.08 %, with the growth rate in Tianjin stood highest at 14.62 %, followed by Hebei Province at 12.49 % and Beijing at 9.51 %.

5.3.1.2 Freight Volume and Freight Turnover

In 2014, freight volume and freight turnover in the Beijing-Tianjin-Hebei region reached 2,915 million tons and 1,666 billion ton-km, respectively, as shown in Table 5.6. For 2006–2014, the average annual growth rate of freight volume in the region was 7.23 %, which was mainly attributed to the high average growth rate of 11.08 % in Hebei Province. During the same period, the growth rate in Tianjin was only 2.16 % while that in Beijing grew negatively.

Table 5.5 Added value of transportation, warehousing and postal service industry of Beijing-Tianjin-Hebei region for 2006–2014 (unit: Billion RMB)

	Beijing	Tianjin	Hebei	Beijing-Tianjin-Hebei region
2006	45.83	25.29	97.15	168.27
2007	50.26	29.41	116.16	195.83
2008	50.57	32.06	128.13	210.76
2009	55.66	47.10	149.19	251.95
2010	64.06	58.52	188.03	310.61
2011	84.53	69.90	204.62	359.05
2012	77.85	72.10	224.11	374.06
2013	88.36	72.51	237.76	398.63
2014	94.81	75.32	249.01	419.14
Average annual growth rate	9.51 %	14.62 %	12.49 %	12.08 %

Source: Compiled from the *Statistical Bulletin of National Economic and Social Development of Beijing, Tianjin and Hebei (2006–2014)*, published by the Bureau of Statistics and the General Research Team of National Bureau of Statistics of Beijing, Tianjin and Hebei

Table 5.6 Freight volume and freight turnover of Beijing-Tianjin-Hebei region for 2014

	Beijing		Tianjin		Hebei	
Index	Value	Growth rate (%)	Value	Growth rate (%)	Value	Growth rate (%)
Freight volume (Million Tons)	295.13	4.30	509.48	−1.30	2110.00	6.00
Railway	11.32	5.00	88.72	5.00	210.00	−8.30
Highway	254.16	3.10	311.30	10.40	1850.00	7.40
Freight turnover (Billion Ton-Kilometers)	67.28	−1.20	335.44	−37.80	1263.14	8.30
Railway	28.44	−12.00	34.90	−3.00	418.60	−1.30
Highway	16.52	5.80	26.52	11.30	701.96	6.70

Source: Compiled from the *Statistical Bulletin of National Economic and Social Development of Beijing, Tianjin and Hebei (2013–2014)*, published by the Bureau of Statistics and the General Research Team of National Bureau of Statistics of Beijing, Tianjin and Hebei

Table 5.7 Ranking of airport throughput and aircraft movements of four major airports in Beijing-Tianjin-Hebei region for 2014

	Ranking of airport throughput in China	Ranking of aircraft movements in China
Beijing Capital International Airport	2	1
Tianjin Binhai International Airport	13	21
ShiJiazhuang Zhengding International Airport	35	41
Beijing Nanyuan Airport	39	49

Source: Compiled from the *Statistical Bulletin of Development of Civil Aviation (2014)*, published by the Civil Aviation Administration of China

5.3.1.3 Airport Throughput

In 2014, airport throughput in Beijing-Tianjin-Hebei region reached 2,164 k-tons, accounting for 15.97 % of the total in China. But airports in the region differ greatly in their air throughput. That in Beijing Capital Airport reached 1,848 k-tons, ranking No. 2 in China; that in Tianjin Airport reached 233.4 k-tons, ranking No. 13 in the country; that in Shijiazhuang Airport and Nanyuan Airport of Hebei stood at 45.6 k-tons and 37.2 k-tons respectively, ranking No. 35 and No. 39 nationwide. Table 5.7 shows the airport throughput and aircraft movements in these four major airports in the region. Besides, the airport throughputs in Handan Airport, Qinhuangdao Shanhaiguan Airport, Tangshan Sannühe Airport and Zhangjiakou Airport were fairly small.

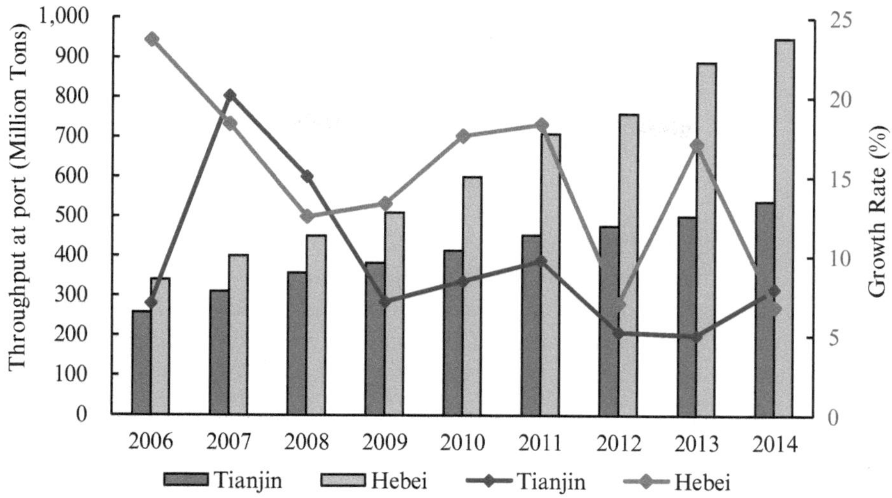

Fig. 5.3 Throughput at port and growth rate of Tianjin and Hebei for 2006–2014
Source: Compiled from the *Statistical Bulletin of National Economic and Social Development of Tianjin and Hebei (2006–2014)*, published by the Bureau of Statistics and the General Research Team of National Bureau of Statistics of Tianjin and Hebei

5.3.1.4 Throughput at Port

The Beijing-Tianjin-Hebei region has four ports whose throughput exceeds 100 million tons, namely, Tianjin Port, Qinhuangdao Port, Tangshan Port and Huanghua Port. The throughput in these ports accounts for about 13.4 % of total port throughput in China. In 2014, throughput in Tianjin Port reached 540 million tons with its container throughput reaching 14.06 million TEUs. The total throughput of the three ports for Hebei Province (Qinhuangdao Port, Tangshan Port, and Huanghua Port) amounted to 950 million tons, as shown in Fig. 5.3.

5.3.2 Transportation Infrastructures

5.3.2.1 Highways

By the end of 2013, highway mileage in Beijing-Tianjin-Hebei region had reached 211 k-km, of which 7,650 km were expressways. The density of highway network and expressway network in the region is 2.33 times and 3.52 times of the respective national average in China. Beijing has eleven national-level expressways including Beijing-Harbin, Beijing-HK-Macao, Beijing-Shanghai, Beijing-Qinhuangdao, Beijing-Kunming, Beijing-Tibet, Beijing-Xinjiang, Beijing-Guangzhou, Beijing-Shijiazhuang, Beijing-Shenyang, and Beijing-Kaifeng expressways. The layout of Tianjin's expressway network is dubbed as "Nine horizontal lines and five vertical lines." The nine horizontal lines refer to Beijing-Qinhuangdao, Beijing-Shenyang, Tangshan-Langfang, State Highway 112, Beijing-Tianjin, Beijing-Tianjin-Tangshan, Tianjin-Shanxi, Tianjin-Baoding, Tianjin-Shijiazhuang, and Nangang expressways. The five vertical lines refer to Beijing-Shanghai, Tangshan-Tianjin, Tanggu-Chengde, Haibin Boulevard and Tianjin-Jixian expressways. Tianjin also has three connecting roads, including Tianjin-Cangzhou, Tianjin-Ninghe, and Beijing-HK expressways. The expressways in operations in Hebei Province include Beijing-Shijiazhuang, Shijiazhuang-Anyang, Shijiazhuang-Taiyuan, Shijiazhuang-Huanghua, Baoding-Tianjin, Tangshan-Tianjin, Beijing-Qinhuangdao, Beijing-Qinhuangdao-Langfang section, Beijing-Shanghai, Xuanhua-Datong, and Beijing-Zhangjiakou expressways. The Qingdao-Yinchuan, Dandong-Lhasa, Beijing-Chengde and Xingtai-Weihai expressways are under construction.

5.3.2.2 Railways

As an important railway hub in China, Beijing has trunk railways such as Beijing-Kowloon, Beijing-Guangzhou, Beijing-Harbin, Beijing-HK-Macau and Beijing-Chengdu railways. Tianjin is not only the junction of two traditional railways (Beijing-Shanghai and Tianjin-Shanhaiguan railways), but also that of other high-speed railways such as Beijing-Shanghai High Speed Railway, Beijing-Tianjin Inter-city Railway, Beijing-Qinhuangdao Passenger Railway, and Tianjin-Baoding Passenger Railway. Tianjin holds an apparent advantage in Continental Bridge transport; it has opened the container trains connecting Tianjin Port to Manzhouli, Alataw Pass and Erenhot, thus becoming the sole eastern originating point linking the Continental Bridge which has three passages in Mainland China (Tianjin Development and Reform Commission 2011). Hebei Province now has 25 trunk railways running through it, including Beijing-Shanghai, Beijing-Guangzhou, Beijing-Kowloon, Beijing-HK-Macau, Shenmu-Huanghua, and Dalian-Qinhuangdao railways.

5.3.2.3 Ports

The Beijing-Tianjin-Hebei region has four ports whose throughput exceeds 100 million tons, and they are Tianjin Port, Qinhuangdao Port, Tangshan Port and Huanghua Port. Because of their large capacity in container transport, these ports occupy significant nodal positions in the regional logistic system.

Tianjin Port is the largest comprehensive trade port in North China. In 2014, throughput of Tianjin Port ranked No. 3 in China with shipping routes connecting to over 500 ports in more than 180 countries and regions in the world. It is the major ocean outlet for the inland countries like Mongolia. Tangshan Port is the important node of transport system in North China for energy, raw materials and other bulk goods. In 2014, throughput of Tangshan Port ranked No. 4 in China, trailing only Tianjin Port in the region. Qinhuangdao Port lies in the east of Liaodong Bay in

Bohai Sea. In 2014, its throughput placed the Port in the 9th place in China. With developed transport facilities, the Port has advantages in container transport. Huanghua Port lies in the junction of Hebei and Shandong Provinces and in the middle part of Circum-Bohai-Sea Economic Circle. Since 2011, its throughput has climbed to more than 100 million tons for four years in a row.

5.3.2.4 Airports

The Beijing-Tianjin-Hebei region has four airports in Beijing, Tianjin, Shijiazhuang and Nanyuan. Beijing Capital International Airport is one of the largest and busiest airports in the world. It has, up to date, opened more than 200 international and domestic routes, playing a significant hub role among the airports in the region. Tianjin Binhai International Airport is a domestic trunk airport, international regular flights airport, grade-I aviation port and major air cargo center. Tianjin Binhai airport advocates the strategy of "highlighting cargo while developing both cargo and passenger transport." Nowadays, it has 86 international and domestic air routes reaching 65 cities, with 1,400 flights weekly. Shijiazhuang Zhengding International Airport is a provincial capital city trunk airport in China. To date, it has opened 47 air routes reaching 47 large and medium-sized airports in China. Nowadays, the operating capacity of Beijing Airport has been saturated while that in Tianjin Airport and Shijiazhuang Airport are in surplus, representing great potential for future growth. As a result, Beijing-Tianjin-Hebei region needs to efficiently allocate its aviation resources, which will also raise higher requirements on coordinated development of air transport in the region (Tianjin Development and Reform Commission 2011).

5.3.3 Development of IT-Based Logistics

The Beijing-Tianjin-Hebei region has established many logistic information platforms, including China Logistic Public Information Platform, Beijing Logistic Public Information Platform, Tianjin Electronic Port Whole-Process Logistic Information Service Platform, Tianjin Cargo Shipping Website, and Hebei Logistic Website. However, Beijing-Tianjin-Hebei region faces the following problems in building its logistic information platforms. First, most of them are cargo allocation, resource and industry vertical portal logistic information platforms. There are few information exchange or application service platforms. Second, because logistic information platforms in Beijing, Tianjin and Hebei Province have different standards, it is difficult to realize effective communications of logistic information and effective allocation of logistic resources.

The Beijing-Tianjin-Hebei ports have made marked progress in IT development. In July 2014, the Beijing-Tianjin-Hebei Customs' Integrated Regional Clearance Reform was launched in Beijing Customs and Tianjin Customs. The reform, by establishing one center and four platforms, can achieve efficient law enforcement and seamless connection of Customs in the three localities. The "one center" refers to the regional Customs clearance center, i.e., while retaining the existing institutions, the originally independent clearance management systems are connected via information network to form one clearance center. The "four platforms" include those for unified declaration, risk control, professional document examination, and field operations (Xinhuanet 2013).

5.3.4 Logistic Enterprises and Logistics Parks

5.3.4.1 Logistic Enterprises

The logistic service industrial chain in Beijing-Tianjin-Hebei region has basically been established, professional logistic enterprises have realized preliminary development, and comprehensive logistic enterprises are gradually becoming major force of modern logistic industry in the region. As per geographical layout, most large logistic enterprises and headquarters of logistic enterprises are congregated in Beijing, whereas few large and notable logistic enterprises or their headquarters are set up in Tianjin or Hebei Province. By February 2, 2015, of the 195 5A logistic enterprises approved by the China Federation of Logistics & Purchasing, 19.5 % are in Beijing-Tianjin-Hebei region, wherein 25 are in Beijing, four are in Tianjin and nine are in Hebei.

Logistic Parks and Centers Beijing has the Tianzhu Free Trade Zone, Beijing Economic Development Zone Bonded & Logistic Center and other logistic parks with bonded functions. It has built a few logistic bases in Shunyi Airport, Tongzhou Majuqiao, Pinggu Mafang and Daxing Jingnan, and other logistic centers such as Shibalidian Logistic Center and Southwest Logistic Center (Beijing Municipal Commission of Commerce and Beijing Municipal Commission of Development and Reform 2011). Along Tianjin Port, a number of logistic parks have been built, including the Tianjin Port Container Center, Tianjin Port Bulk Cargo Center, and Development Zone Bonded & Logistic Center. Adjacent to the Airport, Tianjin has built the Airport International Logistic Park, Binhai CMSTD Composite Logistic Park, and Binhai New Area Comprehensive Bonded Area (Tianjin Development and Reform Commission 2011). In Hebei Province, 19 provincial-level logistic clusters and seven Circum-Beijing modern logistic parks, such as Tangshan Port and Fengrun North, are now under construction.

5.4 Beijing-Tianjin-Hebei Coordinated Development Strategy Engenders Great Impact on Logistic Development

5.4.1 Port Customs Clearance Facilitation Propels Logistic Integration

Ports constitute the core strategic resource of the new starting point for Beijing-Tianjin-Hebei region to develop export-oriented economy and build the Maritime Silk Road. The Customs clearance development in Beijing-Tianjin-Hebei ports helps exploit the port resource advantage of the region, promote inter-connectivity among ports, build service platform for facilitation of investment and trade, and set up a new model for regional port partnership.

In May 2014, the Chinese government issued the *Plan of Reforming Customs Clearance Integration in Beijing-Tianjin-Hebei Customs Region*. In July 2014, the *Reform* was launched in Beijing Customs and Tianjin Customs. In September 2014, Shijiazhuang Customs adopted the integrated regional Customs clearance. The *Reform* thrusts the process of logistic integration in Beijing-Tianjin-Hebei in the following ways. The effects of the *Reform* are: first, trans-regional Customs clearance becoming more convenient and speedy; second, more enterprises are benefited; third, import and export logistics becoming smoother; fourth, Customs service becoming more streamlined (Official website of General Administration of Customs of the People's Republic of China 2014).

5.4.2 Inter-connected Transport Infrastructures Help Improve Transport Convenience

Beijing is the nation's center in railways, highways and air networks. Tianjin is the largest international shipping and logistics center in North China. Hebei boasts abundant port and entry resources. As the Beijing-Tianjin-Hebei integration accelerates, the construction, management and operations of infrastructures in the three regions will galvanize a closer partnership and complementary advantage. They will work in concert to build a hassle-free, rapid and convenient comprehensive transport system while solving the problem of coordinated development by establishing a smoother people and goods flow.

In Beijing-Tianjin-Hebei region, the high-speed railway network has been basically constructed, which consists of Beijing-Tianjin Inter-city Railway, Beijing-Shijiazhuang Passenger Railway, Tianjin-Qinhuangdao High-speed Railway, and Beijing-Shanghai High-speed Railway. It takes one hour or about to travel between major cities in the region such as Beijing-Tianjin, Beijing-Shijiazhuang, Tianjin-Tangshan, Tianjin-Qinhuangdao, and Tianjin-Cangzhou; thus connecting the cities within the "one-hour urban circle." Presently, the region also has several high-speed railways under construction, namely, Tianjin-Baoding Railway, and Beijing-Tianjin Inter-city Railway Extension (to Tianjin Binhai New Area). Other high-speed railways being planned include Beijing-Zhangjiakou Inter-city Railway, Beijing-Chengde Inter-city Railway, Beijing-Tangshan Inter-city Railway, and Tianjin-Shijiazhuang Inter-city Railway. The governments of Beijing, Tianjin and Hebei Province and China Railway Corporation had signed an agreement to establish the Beijing-Tianjin-Hebei Inter-city Railway Investment Co., Ltd. As for highways, the transport network has been established, which includes the newly-completed Beijing-Kunming Expressway and the Tianjin-Shijiazhuang Expressway, which is under construction.

5.4.3 Accelerated Port Integration Drives Resource Integration in Beijing-Tianjin-Hebei Ports

Ports are major engine fueling the logistic development in Beijing-Tianjin-Hebei region. Tianjin Port is a world-class man-made deep-water port, the principal hub port in Circum-Bohai-Sea region, and the primary carrier for building the international shipping center and logistics center in North China. Coastal ports in Hebei include three ports in Qinhuangdao, Tangshan (Jingtang Port Area and Caofeidian Port Area) and Huanghua. Tianjin Port and several coastal ports in Hebei constitute the Tianjin-Hebei Coastal Port Cluster, which bears the responsibility of helping Beijing-Tianjin-Hebei modern metropolis circle and the hinterlands in Northern China and Northwest China to partake in the economic globalization (Enorthnet 2012).

When implementing the Beijing-Tianjin-Hebei Coordinated Development Strategy, cooperation for a win-win ending is the mainline of the regional logistic system development. Ports, as key strategic resource for Beijing-Tianjin-Hebei regional logistical development, become the major starting point to activate regional logistic integration. For example, in August 2014, Tianjin Port Group and Hebei Ports Group established the Bohai Sea Tianjin-Hebei Ports Investment & Development Co., Ltd. By building the partnership mechanism featuring interest sharing, the Company not only helps promote optimal allocation of port resources, and realize more rational division of duties and complementary advantages among ports, but also encourages the ports to better play the radiating role of leading the regional economic growth.

5.4.4 Accelerated Urbanization Expanding Needs for Urban Logistics

The Beijing-Tianjin-Hebei Coordinated Development will foster a rational and orderly development of urban system. According to statistics, China's per capita disposable income of urban residents in 2014 reached 28,800 RMB, which is 2.7

times higher than that of rural residents. The per capita consumer spending by urban residents was 20,000 RMB, a figure 2.4 times higher than that of the rural residents (National Bureau of Statistics of China 2015). As Beijing-Tianjin-Hebei region witnesses accelerated urbanization, the regional purchasing power, consumption ability and market size in the region will be raised considerably, resulting in obvious growth in demand for urban logistic services. At the same time, as the residents' income rises and more consumer goods are attracted to the region, cold-chain logistics, e-commerce logistics and other urban logistics will embrace tremendous opportunities.

5.4.5 *Free Trade Zone Promotes International Logistics in Beijing-Tianjin-Hebei Region*

The Tianjin Pilot Test Free Trade Zone has five major tasks. First, to speed up the transformation of government's functions; second, to expand the openness of investment arena; third, to push forward the transformation and upgrading of trade; fourth, to deepen the openness and innovation of financial industry; and fifth, to push forward and implement the major national strategy of Beijing-Tianjin-Hebei Coordinated Development (Enorthnet 2015).

Development of Tianjin Free Trade Zone will promote the international logistics in the region. According to the *China (Tianjin) Pilot Test Free Trade Zone Master Plan*, the Free Trade Zone will "enhance the functions of international shipping service," such as promoting pooling of shipping factors, working to establish shipping development mechanism and operational model with international competitiveness, and fully leveraging the marine-air interaction between Tianjin Port and Binhai International Airport.

5.5 Summary

In general, the Beijing-Tianjin-Hebei region has relatively strong economic strength. However, the region also shows obvious gaps in its internal development level. Beijing and Tianjin stay far ahead of Hebei Province in terms of GDP, stage of industrial development, urban system and spatial layout, and the extent of export-oriented economy. In recent years, the Beijing-Tianjin-Hebei region has witnessed increasing scale of logistic market, gradually developed transport infrastructures, prominent progress in IT-based logistics, in particular port IT, and rapid growth of logistic enterprises and logistic parks. The strategy of Beijing-Tianjin-Hebei Coordinated Development will have profound and multifaceted impact on the region's logistic development.

References

Beijing Municipal Commission of Commerce and Beijing Municipal Commission of Development and Reform (2011) Logistic development plan of Beijing in the 12th five-year plan period, November 2011

Chinanet (2014) Xi Jinping: break away from intra-regional mentality; Beijing-Tianjin-Hebei region should secure cluster development. Chinanet 28 Feb 2014. http://finance.china.com.cn/news/gnjj/20140228/2220138.shtml

Enorthnet (2012) Overview of Tianjin Port: first-class port in the world. Enorthnet. http://news.enorth.com.cn/system/2012/09/06/009954438.shtml. 6 Sep 2012

Enorthnet (2015) Interview with Liu Jiangang: Tianjin free trade zone, a pioneer in institutional innovations. Enorthnet. http://news.enorth.com.cn/system/2015/04/23/030181585.shtml. 23 Apr 2015

National Bureau of Statistics of China (2015) China statistical bulletin of national economic and social development (2014). National Bureau of Statistics of China, Beijing

Official website of General Administration of Customs of the People's Republic of China (2014) General administration of customs issued plan of reforming customs clearance integration in Beijing-Tianjin-Hebei customs region. Official website of General Administration of Customs of the People's Republic of China. 14 May 2014. http://www.customs.gov.cn/publish/portal0/tab49564/info706219.htm

Tianjin Development and Reform Commission (2011) The 12th five-year plan for Tianjin modern logistic industry. Tianjin, Nov 2011

Xinhuanet (2013) General Administration of Customs, on July 1, kicked off reform of customs clearance integration in Beijing-Tianjin-Hebei Region. Xinhuanet. http://news.xinhuanet.com/fortune/2014-05/14/c_1110683295.htm. 14 May 2014

Chapter 6
Logistic Development Along the Yangtze River Economic Belt

Ya Xu

The Yangtze River Economic Belt runs across Eastern China, Central China, and Western China, covering the coastal areas in the East and the vast inland areas, which possesses unique location advantages and huge development potential. The Chinese government has made the significant strategic decision to push forward the development of the Yangtze River Economic Belt (hereinafter referred to as the Belt), while building the new areas that support the growth of Chinese economy. It is of great significance for China to effectively expand the domestic demand, promote stable economic growth, adjust the regional structure, and upgrade the economy. Along the Belt, there are densely dotted industries, rapid growth of logistic needs, well-developed logistic infrastructures, strong support for policies, and rapidly evolving trend of modern logistics. Relying on the Golden Waterways in Yangtze River, shipping logistics provides important support for the development of logistics industry in the Belt, thus greatly promoting the economic growth and formation of industrial belt along the Belt.

This chapter has three sections. Section 6.1 presents an overview of the economic development along the Belt, including major economic indicators, industries, and characteristics of industrial development. Section 6.2 elaborates on the development of the logistic market along the Belt, covering the demand and supply of the logistic market. Section 6.3 explains the development of shipping logistics in the Yangtze River, including the basic conditions and the development features of shipping logistics in the Yangtze River.

Y. Xu (✉)
Business School, Nankai University, Tianjin, People's Republic of China
e-mail: xuya.1981@qq.com

L. Wang et al. (eds.), *Contemporary Logistics in China*, Current Chinese Economic Report Series, DOI 10.1007/978-981-10-1052-1_6

6.1 Overview of Economic Development Along the Yangtze River Economic Belt

The Yangtze River Economic Belt[1] is the economic region in China that has the largest economic aggregate. In recent years, with the Belt being hoisted to the national strategic position, regions along the Yangtze River have grasped the historic opportunity, taken full advantages in location and policies, sped up the economic transformation and upgrading, boosted the economic aggregate, optimized the industrial structure, and enhanced the comprehensive economic strength. Figure 6.1 shows a map of the Belt.

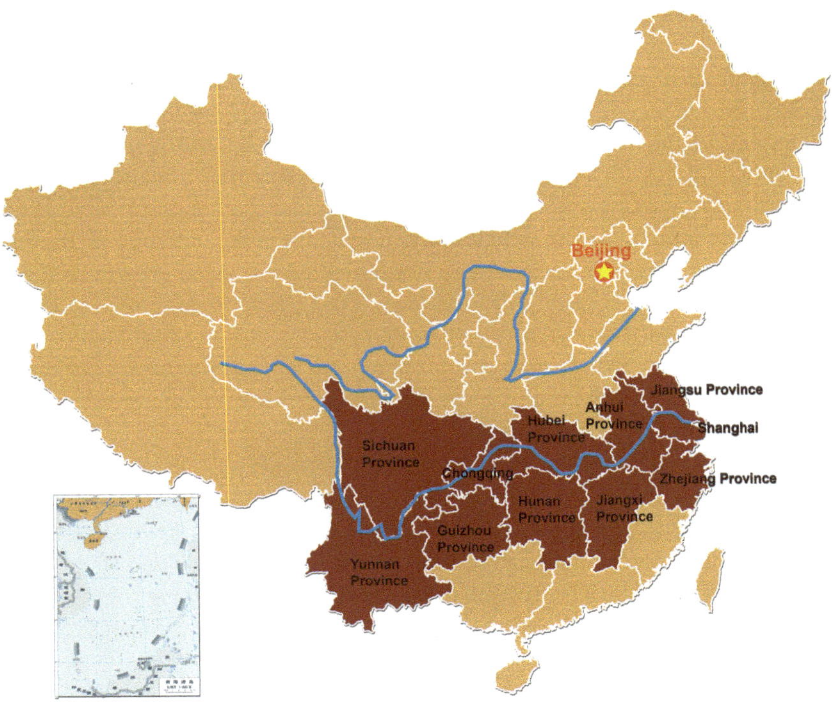

Fig. 6.1 Map of Yangtze River Economic Belt

[1] The Yangtze River Economic Belt starts with Shanghai in the East and reaches Yunnan in the West. The River runs across eleven provinces and municipalities including Shanghai, Jiangsu, Zhejiang, Anhui, Jiangxi, Hubei, Hunan, Chongqing, Sichuan, Guizhou and Yunnan.

6.1.1 Background of Yangtze River Economic Belt Development

The Yangtze River Economic Belt is a massive economic belt featuring abundant resources, numerous economic clusters and densely placed cities. It is also one of the regions in the world with most dynamic economic development and a highly promising prospect. It is regarded as China's 4th Growth Pole following the Yangtze River Delta, the Pearl River Delta and the Circum-Bohai-Sea Region, providing strong support to the economic growth in China.

6.1.1.1 Huge Potential for Regional Cooperation and Development

The population and total GDP in the Yangtze River Economic Belt exceed 40 % of the respective total in China. It has become one of the regions in China with strongest comprehensive strength and greatest strategic support to the economy.

Upstream regions along the Yangtze River feature dense population. Population in Sichuan and Chongqing alone exceeds 100 million people, providing a vast market for enterprises in the middle and lower reaches of the Yangtze River. Enhancing the cooperation among them can foster great benefits.

As the forefront of reform and opening-up in China, the downstream reaches along the Belt need to participate in the higher-level international competition. The upstream regions along the Belt, Yunnan and Chongqing in particular, can help China build closer relations with South Asia and Southeast Asia. Relying on the major transport passage of Chongqing-Europe International Railway, China can enhance its economic contacts between the Belt and the EU nations.

The Yangtze River has abundant water resources, favorable agricultural conditions, dense and closely-connected industry and cities. Given these advantages, if China could promote a well-organized industrial transfer/upgrade, optimize its industries, promote the clustered development of new-type of urban areas, and realize the integration of upstream and downstream reaches, it would be able to establish the powerful force that stimulates the development of a region which accounts for 20 % of the territory and about 600 million people (Wang Rengui 2014).

6.1.1.2 Strategic Position Keeps Advancing

It took about 30 years for the Yangtze River Economic Belt to move up from a conceptual idea to becoming a strategic decision. At present, the conditions for its implementation have matured. On the one hand, China has resolved to build the all-around well-off society by 2020. To realize this goal, one crucial task is to ensure that the trailing Central and Western regions also develop concurrently so that they are in step with the rest of the country. On the other hand, China is now in a special period facing a slow-down in economic development, adjustment of industrial structure, and digestion of the effects of prior stimulus policies. The country is in urgent need of new development impetus so as to promote the economic upgrading and transformation. Nowadays, development in the coastal areas in Eastern China is relatively saturated while development in inland areas of Central and Western China needs to be accelerated. The Belt that connects the East and the West has become the carrier for development in these two regions (Che Haigang 2014).

On November 27, 2005, Shanghai, Jiangsu, Anhui, Jiangxi, Hubei, Hunan, Chongqing, Sichuan and Yunnan signed the Yangtze River Economic Belt Cooperation Agreement in Beijing. Spearheaded by the Ministry of Transport, the signing of the Agreement has affirmed the strategic layout of collaborated development in the Belt.

In September 2013, the National Development and Reform Commission (NDRC) and the Ministry of Transport convened the meeting for drafting the *Guiding Opinions on Building the New Supporting Belt for the Chinese Economy Basing on the Yangtze River*. In addition to the above-mentioned seven provinces and two municipalities, Zhejiang and Guizhou were also included in the program, making up the eleven provinces and municipalities. During the NPC and CPPCC Session in 2014, the Government Work Report, for the first time, placed the Yangtze River Economic Belt into the State Council Work Agenda, signifying that the development of the Belt has been elevated to the level of national strategy.

In June 2014, Premier Li Keqiang convened the Executive Meeting of the State Council to charge the development of comprehensive and vertical transport corridor as well as the Yangtze River Economic Belt. In September 2014, the State Council released the *Guiding Opinions on Pushing forward Development of Yangtze River Economic Belt Basing on the Golden Waterways*, with the *Plan of Comprehensive Vertical Transport Corridor for the Yangtze River Economic Belt (2014–2020)* as the Appendix. These two documents ascertained the strategic position of the Belt as: the inland river economic belt with global influence, the coordinated development belt featuring collaboration and interaction among the Eastern, Central and Western China, the domestically and internationally open belt basing on the coast, the river and bordering areas, and the frontier demonstration belt for ecological development.

6.1.2 Major Economic Indicators of Yangtze River Economic Belt

6.1.2.1 Regional GDP

Economic aggregate of the Yangtze River Economic Belt accounts for nearly 50 % of the total GDP in China. In 2014, the Belt maintained a relatively rapid economic growth. Apart from Shanghai, ten of the eleven provinces and municipalities had a growth rate higher than the nation's average. Among them, the growth rate of Chongqing and Guizhou each exceeded 10 %; that of Jiangsu, Jiangxi, Anhui, Hubei, Hunan, Sichuan and Yunnan each exceeded 8 %. Table 6.1 exhibits the GDP and GDP growth rate of the Yangtze River Economic Belt for 2010–2014.

Table 6.1 GDP and GDP growth rate in provinces and municipalities of the Yangtze River Economic Belt for 2010–2014

Region	2010		2011		2012		2013		2014	
	Value (billion)	Growth rate (%)	Value (billion)	Growth rate (%)	Value (billion)	Growth rate (%)	Value (billion)	Growth rate (%)	Value (billion)	Growth rate (%)
Shanghai	1716.6	10.3	1919.6	8.2	2018.2	7.5	2181.8	7.7	2356.1	7.0
Jiangsu	4090.3	12.6	4860.4	11.7	5405.8	10.1	5916.2	9.6	6508.8	8.7
Zhejiang	2772.2	11.9	3231.9	9.0	3460.6	8.0	3775.7	8.2	4015.4	7.6
Anhui	1226.3	14.5	1511.0	13.5	1721.2	12.1	1903.9	10.4	2084.9	9.2
Jiangxi	945.1	14.0	1170.3	12.5	1294.9	11.0	1433.9	10.1	1570.9	9.7
Hubei	1580.6	14.8	1959.4	13.8	2225.0	11.3	2466.8	10.1	2736.7	9.7
Hunan	1603.8	14.6	1967.0	12.8	2215.4	11.3	2462.2	10.1	2704.9	9.5
Chongqing	792.6	17.1	1001.1	16.4	1141.0	13.6	1278.3	12.3	1426.5	10.9
Sichuan	1689.9	15.1	2102.7	15.0	2385.0	12.6	2626.1	10.0	2853.7	8.5
Guizhou	459.4	12.8	570.2	15.0	680.2	13.6	800.7	12.5	925.1	10.8
Yunnan	722.4	12.3	889.3	13.7	1030.9	13.0	1172.1	12.1	1281.5	8.1
Yangtze River Economic Belt	17,599.2	20.6	21,182.9	20.4	23,578.2	11.3	26,017.6	10.3	28,464.4	9.4

Note: Growth rate for the Yangtze River Economic Belt is nominal growth rate; all others are actual growth rate.
Source: Compiled from the *Statistical Bulletin of National Economic and Social Development* (*2010–2014*) of Shanghai, Jiangsu Province, Zhejiang Province, Anhui Province, Jiangxi Province, Hubei Province, Hunan Province, Chongqing, Sichuan Province, Guizhou Province and Yunnan Province, published by the National Bureau of Statistics of Shanghai, Jiangsu Province, Zhejiang Province, Anhui Province, Jiangxi Province, Hubei Province, Hunan Province, Chongqing, Sichuan Province, Guizhou Province and Yunnan Province

6.1.2.2 Industrial Added Value

During 2010–2014, the industrial added value of the Yangtze River Economic Belt rose rapidly with a five year average annual growth rate reaching 11.4 %. In 2014, the industrial added value of the Belt amounted to 11.5 trillion RMB, amounting to 50 % of the total in China. The growth rate of industrial added value in the Belt reached 7.8 %, which is 1.1 times more than the national average in China. The industrial added value in five major industrial provinces, namely, Jiangsu, Zhejiang, Hubei, Hunan and Sichuan, accounted for 68.8 % of the total in the Belt. Table 6.2 shows the industrial added value and growth rate of the Belt for 2010–2014.

6.1.2.3 Total Retail Sales of Consumer Goods

During 2010–2014, total retail sales of consumer goods of the Yangtze River Economic Belt rose rapidly with an average annual growth rate reaching 14.2 %. In 2014, the total retail sales of consumer goods valued at 10.7 trillion RMB, accounting for 40 % of the total in China. Table 6.3 gives the total retail sales of consumer goods and growth rate of the Yangtze River Economic Belt for 2010–2014.

6.1.2.4 Total Value of Import and Export Trade

During 2010–2014, import and export volume of Yangtze River Economic Belt recorded an average annual growth rate of 9.1 %. In 2014, total volume of import and export of the Belt reached 1,757.46 billion USD, accounting for 40 % of the total in China; the growth rate of import and export volume of the Belt reached 7.2 %, or 3.1 times more than the national average. Shanghai, Jiangsu and Zhejiang have developed an export-oriented economy. In 2014, their combined import and export volume accounted for 78.8 % of the total of the Belt, equaling 32.6 % of the total for China. Table 6.4 shows the import and export volume in the provinces and municipalities of the Yangtze River Economic Belt for 2010–2014.

6.1.3 Industrial Development in the Yangtze River Economic Belt

6.1.3.1 Industrial Structure of the Yangtze River Economic Belt

During 2010–2014, the Yangtze River Economic Belt methodically pushed forward industrial transfer, optimization, and upgrading, resulting in significant outcome. In general, modern agriculture in the Belt developed steadily, machinery manufacturing developed towards the large-scale and high-end trend, and high-tech industry and modern service industry expanded swiftly. In 2014, for nine provinces and

Table 6.2 Industrial added value and growth rate in provinces and municipalities of the Yangtze River Economic Belt for 2010–2014

Region	2010		2011		2012		2013		2014	
	Added value (billion)	Growth rate (%)	Added value (billion)	Growth rate (%)	Added value (billion)	Growth rate (%)	Added value (billion)	Growth rate (%)	Added value (billion)	Growth rate (%)
Shanghai	645.7	17.5	723.1	7.5	715.9	2.8	723.7	6.3	736.3	4.3
Jiangsu	2122.4	16.0	2503.5	13.8	2660.6	12.6	2939.9	11.3	3150.8	7.5
Zhejiang	1039.7	16.2	1087.8	10.9	1087.5	7.1	1170.1	8.5	1254.3	6.9
Anhui	560.2	23.6	706.2	21.1	755.1	16.2	858.5	13.7	954.6	11.2
Jiangxi	428.7	19.9	541.2	17.3	582.8	13.4	643.4	11.9	699.5	11.2
Hubei	613.7	23.6	856.6	20.5	955.2	14.6	1068.0	11.8	1183.3	10.8
Hunan	630.5	21.2	812.3	18.2	913.9	13.5	1000.1	11.1	1075.0	9.2
Chongqing	291.7	17.4	369.8	22.9	469.0	22.2	498.1	15.9	525.0	13.1
Sichuan	732.6	22.9	949.1	21.6	1080.1	15.6	1157.9	11.0	1240.9	9.4
Guizhou	135.4	15.8	172.4	21.0	205.5	21.3	253.2	13.1	311.8	11.3
Yunnan	260.6	14.7	320.6	21.0	345.1	16.2	376.8	13.6	389.9	11.3
Yangtze River Economic Belt	7461.2	23.4	9042.4	21.2	9770.7	8.1	10,689.6	9.4	11,521.3	7.8

Note: 1. Growth rate for the Yangtze River Economic Belt is nominal growth rate; all others are actual growth rate.
2. The industrial added value in Zhejiang, Anhui, Hubei and Guizhou Province refers to those generated by enterprises above designated size.
Source: Statistical Bulletin of National Economic and Social Development (2010–2014) of Shanghai, Jiangsu Province, Zhejiang Province, Anhui Province, Jiangxi Province, Hubei Province, Hunan Province, Chongqing, Sichuan Province, Guizhou Province and Yunnan Province

Table 6.3 Total retail sales of consumer goods and growth rate of the Yangtze River Economic Belt for 2010–2014

Region	2010		2011		2012		2013		2014	
	Value (billion RMB)	Growth rate (%)	Value (billion RMB)	Growth rate (%)	Value (billion RMB)	Growth rate (%)	Value (billion RMB)	Growth rate (%)	Value (billion RMB)	Growth rate (%)
Shanghai	603.7	17.5	677.7	12.3	738.7	9.0	801.9	8.6	871.9	8.7
Jiangsu	1348.2	18.7	1584.2	17.5	1821.5	15.0	2065.7	13.4	2320.9	12.4
Zhejiang	1024.5	18.8	1202.8	17.4	1358.8	13.0	1522.6	12.0	1690.5	11.7
Anhui	415.2	19.2	490.1	18.0	568.6	16.0	648.1	14.0	732.1	13.0
Jiangxi	295.6	19.0	348.5	17.9	402.7	15.6	455.1	13.6	512.9	12.7
Hubei	671.9	19.0	792.8	18.0	919.7	16.0	1046.6	13.8	1180.6	12.8
Hunan	577.5	19.1	680.9	17.9	785.5	15.4	894.1	13.8	1008.2	12.8
Chongqing	293.9	18.5	348.8	18.7	403.4	15.7	451.2	14.0	509.6	13.0
Sichuan	663.5	18.7	783.7	18.1	908.8	16.0	1035.5	13.9	1166.6	12.7
Guizhou	148.3	18.9	172.9	18.1	200.5	16.0	236.6	14.0	258.0	12.9
Yunnan	250.0	21.9	300.0	20.0	354.2	18.0	403.6	14.0	463.3	12.7
Yangtze River Economic Belt	6292.3	17.2	7382.4	17.3	8462.4	14.6	9561.0	13.0	10,714.6	12.1

Note: Growth rates in the table are nominal growth rates.
Source: Statistical Bulletin of National Economic and Social Development (*2010–2014*) of Shanghai, Jiangsu Province, Zhejiang Province, Anhui Province, Jiangxi Province, Hubei Province, Hunan Province, Chongqing, Sichuan Province, Guizhou Province and Yunnan Province

Table 6.4 Import and export volume in the provinces and municipalities of the Yangtze River Economic Belt for 2010–2014

Region	2010		2011		2012		2013		2014	
	Value (billion USD)	Growth rate (%)	Value (billion USD)	Growth rate (%)	Value (billion USD)	Growth rate (%)	Value (billion USD)	Growth rate (%)	Value (billion USD)	Growth rate (%)
Shanghai	368.9	32.8	437.4	18.6	436.8	–0.2	441.4	1.1	466.6	5.6
Jiangsu	465.8	37.5	539.8	15.9	548.0	1.5	550.8	0.5	563.8	2.3
Zhejiang	253.5	35.0	309.4	22.0	312.2	0.9	335.8	7.5	355.2	5.8
Anhui	24.3	54.8	31.3	29.1	39.3	25.6	45.6	16.2	49.3	8.2
Jiangxi	21.5	67.9	31.6	46.1	33.4	6.2	36.7	10.0	42.8	16.4
Hubei	25.9	50.2	33.5	29.1	32.0	4.8	36.4	13.8	43.1	18.4
Hunan	14.7	44.7	19.0	29.6	21.9	15.5	25.2	14.7	30.8	22.2
Chongqing	12.4	61.1	29.2	140.0	53.2	82.8	68.7	29.1	95.5	39.0
Sichuan	32.8	35.6	47.8	46.2	59.1	23.9	64.6	9.2	70.3	8.8
Guizhou	3.1	36.0	4.9	55.2	6.6	35.7	8.3	25.0	10.7	29.5
Yunnan	13.4	66.5	16.1	19.6	21.0	31.0	25.8	22.9	29.6	17.1
Yangtze River Economic Belt	1236.2	37.0	1499.9	21.3	1563.6	4.2	1639.4	4.8	1757.5	7.2

Note: Growth rates in the table are nominal growth rates.
Source: Statistical Bulletin of National Economic and Social Development (*2010–2014*) of Shanghai, Jiangsu Province, Zhejiang Province, Anhui Province, Jiangxi Province, Hubei Province, Hunan Province, Chongqing, Sichuan Province, Guizhou Province and Yunnan Province

Table 6.5 Proportion of tertiary industry in provinces and municipalities of the Yangtze River Economic Belt for 2010–2014

				Unit: %		
	2010			2014		
Region	Primary	Secondary	Tertiary	Primary	Secondary	Tertiary
Shanghai	0.7	42.3	57	0.5	34.7	64.8
Jiangsu	6.2	53.2	40.6	5.6	47.7	46.7
Zhejiang	6.7	53.4	39.9	4.4	47.7	47.9
Anhui	14.1	52.1	33.8	11.5	53.7	34.8
Jiangxi	10.2	46.8	43	10.7	53.4	35.9
Hubei	13.6	49.1	37.3	11.6	46.9	41.5
Hunan	14.7	46	39.3	11.6	46.2	42.2
Chongqing	8.7	55.2	36.1	7.4	45.8	46.8
Sichuan	14.7	50.7	34.6	12.4	50.9	36.7
Guizhou	13.7	39.2	47.1	13.8	41.6	44.6
Yunnan	15.3	44.7	40	15.5	41.2	43.2
Yangtze River Economic Belt	9.2	50.4	40.4	8.4	46.9	44.7

Source: Statistical Bulletin of National Economic and Social Development (2010–2014) of Shanghai, Jiangsu Province, Zhejiang Province, Anhui Province, Jiangxi Province, Hubei Province, Hunan Province, Chongqing, Sichuan Province, Guizhou Province and Yunnan Province

municipalities out of the eleven in the Belt, the proportion of the tertiary industry rose considerably from that in 2010. Table 6.5 shows the proportion of tertiary industry in provinces and municipalities of the Yangtze River Economic Belt for 2010–2014.

6.1.3.2 Main Characteristics of Industrial Development

Size and Capacity of Industry Hold Weighty Position in China

The Yangtze River Economic Belt is a major industrial region in China which has modern industries including iron and steel, auto, electronics, and petro-chemicals. According to rough statistics, there are presently five iron and steel producers in the Belt, i.e., Wuhan Iron and Steel, Bao Steel, Panzhihua Iron and Steel, Nanjing Steel, and Ma Steel, about ten refinery plants and 400,000 chemical enterprises. In Jiangsu Province alone, there are eight port-bordering chemical areas with more than 100,000 chemical enterprises. Along the Yangtze River, production of steel and iron accounts for 36 % of the total in China; production volume of autos and petro-chemicals each exceeds 40 % of the respective total in China. Even more, many high-tech industries and super-large enterprises are also concentrated along the Belt (Zhou Yanhua 2014).

Till today, the Belt has established various industrial clusters pillared on auto, iron and steel, pharmaceuticals, petro-chemicals, silk, electronics, finance and IT industry, pooling a large number of leading and powerful enterprises in these fields. Because of the solid industrial foundation, the Belt has become a major region sustaining the economic aggregate in China. In 2014, the industrial added value in the Belt reached 11.486 trillion RMB, accounting for 50.38 % of the total in China; the industrial profits stood at 2,816.83 billion RMB, representing 43.53 % of the total in China. Besides, 36.5 % of the electric power in China is supplied by the Belt.

High-Tech Industry Develops Rapidly

Yangtze River Economic Belt is the dynamic region of high-tech development. The 17 national-level high-tech development zones and a number of provincial and municipal-level development zones are stretched out in Shanghai, Nanjiang, Suzhou, Hefei, Jiujiang, Nanchang, Wuhan, Huangshi, Changsha and Chongqing. In 2014, the high-tech project contracts in the Belt amounted to total worth of 329.21 billion RMB, accounting for 38.38 % of the total in China. The average growth rate of total contractual worth of high-tech projects in the Belt is 18.82 %, which is 4 percentage points higher than the national average. Due to the high-tech stimulus, the GDP growth rate in the Belt is higher than the national average. In addition, over the past five years, the Internet industry in the Belt has been booming rapidly; a large number of outstanding Internet enterprises have emerged, including Alibaba in Hangzhou and YHD in Shanghai.

Gradient Transfer of Industry Is Accelerated

Yangtze River Economic Belt is the region in China with most evident gradient development. From the East to the West of the region, there are areas in different stages of development. As industrial restructuring in the Belt advancing, and when developed regions in Eastern China are being constrained by limited resources, higher labor cost, decaying environment and market conditions, many industries began migrating to the Central and Western regions in succession. After many years of capital input and market reform, relevant transfer mechanisms and institutions have been improved in Central and Western China; they are better suited to take in the industrial transfer from Eastern China. For instance, during 2011–2013, Chongqing brought in 1,162.3 billion RMB of capital investment from coastal areas in Eastern China; it carried out 19,743 projects each with investment of over ten million RMB, meaning that, on average, 18 projects were brought in to Chongqing every day. The city has achieved evident results in receiving the industrial transfer. It has fostered and established a number of industrial clusters such as electronic parts and components, auto parts and components, refined petro-chemicals, textile, apparel and consumer goods manufacturing. The manufacturing capacity it has received amounted to 800 billion RMB.

6.1.3.3 Mainstay Industries in Various Regions of the Yangtze River Economic Belt

The upstream, mid-stream and downstream regions of the Belt differ greatly in stages of economic development and industrial structure. The Yangtze River Delta in the downstream reaches features well-developed economy and strong technological capability. It is where the high-tech R&D centers, financial and trade centers, high-tech industries, and headquarters congregate. The mid-stream region along the Yangtze River has a solid foundation in machinery manufacturing with strong auxiliary industrial capability. It is the major industrial base in China. The upstream region possesses abundant resources but has a relatively less-developed economy focusing on resource industries. Based on their own characteristics and development needs, the provinces and municipalities along the Belt have actively optimized and upgraded their respective industries, thus developed their own distinctive industries. Table 6.6 lists the mainstay industries in provinces and municipalities in the Belt.

Table 6.6 Mainstay industries in provinces and municipalities in the Yangtze River Economic Belt

Region	Mainstay industries
Shanghai	PC, telecom and other electronic device manufacturing; auto manufacturing; electrical machinery and apparatus manufacturing; chemical raw materials and chemicals manufacturing; petro processing, coking and nuclear fuel processing; black metal smelting and rolling processing; power, heat production and supply
Jiangsu	Auto manufacturing; pharmaceuticals production; special equipment manufacturing; electrical machinery and apparatus manufacturing; PC, telecom and other electronic device manufacturing; building industry
Zhejiang	Textile; chemical raw materials and chemicals manufacturing; general equipment manufacturing; electrical machinery and apparatus manufacturing; power, heat production and supply; rubber and plastic production
Anhui	Electric machinery and apparatus manufacturing; coal mining and sorting; agricultural and food processing; chemical raw materials and chemicals manufacturing; non-metal mineral production; black metal smelting and rolling processing; general equipment manufacturing and auto manufacturing
Jiangxi	Agricultural and food processing; chemical raw materials and chemicals manufacturing; non-metal mineral production; non-ferrous metal smelting and rolling processing; electrical machinery and apparatus manufacturing; black metal smelting and rolling processing
Hubei	Agricultural and food processing; auto manufacturing; auto manufacturing; textile; electric machinery and apparatus manufacturing; PC, telecom and other electronic device manufacturing; black metal smelting and rolling processing; non-metal mineral production; chemical raw materials and chemicals manufacturing; power, heat production and supply
Hunan	Agricultural and food processing; chemical raw materials and chemicals manufacturing; special equipment manufacturing; non-ferrous metal smelting and rolling processing; non-metal mineral production; and black metal smelting and rolling processing

(continued)

Table 6.6 (continued)

Region	Mainstay industries
Chongqing	Agricultural and food processing; non-metal mineral production; black metal smelting and rolling processing; non-ferrous metal smelting and rolling processing; auto manufacturing; railway, ship, aviation, space and other transport equipment manufacturing; PC, telecom and other electronic device manufacturing; electrical machinery and apparatus manufacturing
Sichuan	Agricultural and food processing; liquor, beverages and tea production; chemical raw materials and chemicals manufacturing; non-metal mineral production; black metal smelting and rolling processing; PC, telecom and other electronic device manufacturing
Guizhou	Coal mining and sorting; liquor, beverages and tea production; tobacco production; power, heat production and supply
Yunnan	Tobacco production; power, heat production and supply; non-ferrous metal smelting and rolling processing

Source: Statistical Bulletin of National Economic and Social Development (2014) of Shanghai, Jiangsu Province, Zhejiang Province, Anhui Province, Jiangxi Province, Hubei Province, Hunan Province, Chongqing, Sichuan Province, Guizhou Province and Yunnan Province

6.2 Development of Logistics Market in the Yangtze River Economic Belt

Founded on the solid industrial bases, propitious policy environment and well-developed comprehensive transportation system, the logistic industry in the Yangtze River Economic Belt has been developing rapidly. Favorable conditions include continually expanding logistic market, congregation of large and celebrated logistic enterprises, continuously improved logistic infrastructures, rapidly elevated logistic informatization level, and gradually rising logistic supply capacity and service level year by year.

6.2.1 Logistics Demand of Yangtze River Economic Belt

6.2.1.1 Added Value of Transportation, Storage and Postal Service

In 2013, added value of transportation, warehousing and postal service amounted to 1,071.4 billion RMB, accounting for 40 % of the nation's total, at a year-on-year growth rate at 7.9 %. Table 6.7 gives the added value of transportation, warehousing and postal service of the provinces in the Yangtze River Economic Belt for 2013.

6.2.1.2 Freight Volume and Freight Turnover

In 2013, the freight volume and turnover in the Yangtze River Economic Belt reached 18.2 billion tons and 6,944.8 billion tons-km, growing by 8.2 % and 8.6 %, respectively. During 2009–2013, the overall freight volume and turnover in the Belt

Table 6.7 Added value of transportation, warehousing and postal service of the provinces in the Yangtze River Economic Belt for 2013

Region	Added value (billion RMB)	Growth rate (%)	Proportion in local GDP (%)
Shanghai	93.5	1.0	4.3
Jiangsu	253.0	7.6	4.3
Zhejiang	132.6	4.7	3.5
Anhui	70.7	7.2	3.7
Jiangxi	67.9	7.6	4.7
Hubei	107.8	10.5	4.4
Hunan	107.8	11.4	4.9
Chongqing	58.1	9.2	4.6
Sichuan	75.2	6.3	2.9
Guizhou	77.5	12.7	9.7
Yunnan	27.4	9.1	2.3
Yangtze River Economic Belt	1071.4	7.9	4.5

Source: Compiled from the *Statistical Bulletin of National Economic and Social Development (2013)* of Shanghai, Zhejiang Province and Yunnan Province, the *Statistical Yearbook (2013)* of Jiangsu Province, Anhui Province, Jiangxi Province, Hubei Province, Hunan Province, Chongqing, Sichuan Province and Guizhou Province, published by the National Bureau of Statistics of Shanghai, Jiangsu Province, Zhejiang Province, Anhui Province, Jiangxi Province, Hubei Province, Hunan Province, Chongqing, Sichuan Province, Guizhou Province and Yunnan Province

grew considerably, but with a declining growth rate. Tables 6.8 and 6.9 present the freight volume and turnover in the Yangtze River Economic Belt for 2009–2013.

6.2.1.3 Cargo Throughput at Port

In 2009–2013, cargo throughput at port in the Yangtze River Economic Belt rose rapidly. In 2013, cargo throughput at port in the Belt reached 5.66 billion tons, rising by 7.5 %. In 2013, the cargo throughput at port in Jiangsu Province and Zhejiang Province amounted to 2.14 billion tons and 1.38 billion tons, respectively. The total of these two provinces accounted for 62 % of the cargo throughput at port in the Belt. Table 6.10 lists the cargo throughput at port and the growth rate of the provinces and municipalities in the Yangtze River Belt for 2009–2013.

6.2.1.4 Airport Cargo Throughput

The airport cargo throughput in the Yangtze River Economic Belt mainly concentrates in large and medium-sized hub airports such as Shanghai Pudong, Shanghai Hongqiao, Chengdu Shuangliu and Hangzhou Xiaoshan airports. In 2014, the cargo throughput in these four airports ranked No. 1, No. 6, No. 5 and No. 7 in China. Shanghai Pudong Airport is one of the three airport hubs in the country; its cargo

Table 6.8 Freight volume in the Yangtze River Economic Belt for 2009–2013

Province/municipality	2009		2010		2011		2012		2013	
	Freight volume (million tons)	Growth rate (%)	Freight volume (million tons)	Growth rate (%)	Freight volume (million tons)	Growth rate (%)	Freight volume (million tons)	Growth rate (%)	Freight volume (million tons)	Growth rate (%)
Shanghai	769.7	–8.8	810.2	5.3	933.2	15.2	943.8	1.1	915.4	–3.0
Jiangsu	1609.7	9.2	1884.0	17.0	2125.9	12.8	2313.0	8.8	2516.9	8.8
Zhejiang	1512.4	3.1	1705.4	12.8	1856.9	8.9	1910.6	2.9	1878.9	–1.7
Anhui	1970.0	10.1	2281.0	16.0	2680.0	17.7	3120.0	16.1	3570.0	15.6
Jiangxi	856.3	6.7	1003.2	17.0	1115.7	11.2	1270.2	13.8	1404.0	10.5
Hubei	827.1	9.2	970.1	17.3	1101.7	13.6	1253.9	13.8	1397.4	11.4
Hunan	1288.9	11.0	1498.4	16.1	1687.7	12.6	1913.8	13.4	2114.1	10.5
Chongqing	684.9	7.6	813.8	18.8	967.8	18.9	1101.4	13.8	974.0	12.7
Sichuan	1180.9	3.1	1333.6	12.9	1538.3	15.3	1744.5	13.4	1896.1	8.7
Guizhou	348.0	6.5	397.4	14.2	448.9	13.0	526.5	17.3	727.0	38.1
Yunnan	474.0	3.3	528.0	11.2	668.0	15.4	759.0	13.6	842.0	11.0
Yangtze River Economic Belt	11,522.0	6.5	13,225.2	14.8	15,124.1	14.4	16,856.6	11.5	18,235.8	8.2

Source: Compiled from the *Statistical Bulletin of National Economic and Social Development* (2009–*2013*) of Shanghai, Jiangsu Province, Anhui Province, Jiangxi Province, Hunan Province, Chongqing, Guizhou Province and Yunnan Province, the *Statistical Yearbook* (*2013*) of Zhejiang Province, Hubei Province and Sichuan Province, published by the National Bureau of Statistics of Shanghai, Jiangsu Province, Zhejiang Province, Anhui Province, Jiangxi Province, Hubei Province, Hunan Province, Chongqing, Sichuan Province, Guizhou Province and Yunnan Province

Table 6.9 Freight turnover in the Yangtze River Economic Belt for 2009–2013

	2009		2010		2011		2012		2013	
Province/municipality	Freight turnover (billion tons-km)	Growth rate (%)	Freight turnover (billion tons-km)	Growth rate (%)	Freight turnover (billion tons-km)	Growth rate (%)	Freight turnover (billion tons-km)	Growth rate (%)	Freight turnover (billion tons-km)	Growth rate (%)
Shanghai	1443.6	−9.9	1617.3	12.0	2037.6	26.0	2042.7	0.3	1786.8	−12.5
Jiangsu	515.5	8.5	611.2	18.6	751.4	22.9	847.5	12.8	950.5	12.2
Zhejiang	491.9	0.7	711.2	25.7	862.7	21.3	918.3	6.4	986.7	7.4
Anhui	627.3	7.5	714.4	13.1	843.5	18.1	979.3	15.9	1113.7	14.1
Jiangxi	235.1	2.3	273.6	16.4	300.4	9.7	344.9	14.8	390.4	13.2
Hubei	280.8	4.1	336.8	20.4	404.4	21.2	469.4	16.1	488.3	12.2
Hunan	253.8	5.6	295.8	16.3	340.2	15.1	400.7	12.8	428.2	6.9
Chongqing	164.4	10.6	201.0	22.3	253.0	25.9	264.8	4.7	300.0	13.3
Sichuan	152.6	0.8	171.0	12.3	190.9	11.6	213.0	11.6	243.8	8.1
Guizhou	92.6	15.0	100.6	8.6	106.1	5.4	117.5	10.7	129.5	10.2
Yunnan	90.4	3.9	99.1	8.8	107.0	8.0	116.5	8.9	127.1	9.2
Yangtze River Economic Belt	4348.0	5.4	5131.9	15.9	6197.3	16.8	6714.5	10.5	6944.8	8.6

Source: Compiled from the *Statistical Bulletin of National Economic and Social Development* (*2009–2013*) of Jiangsu Province, Zhejiang Province, Anhui Province, Jiangxi Province, Hubei Province, Hunan Province, Sichuan Province, Guizhou Province and Yunnan Province, the *Statistical Yearbook* (*2013*) of Shanghai and Chongqing, published by the National Bureau of Statistics of Shanghai, Jiangsu Province, Zhejiang Province, Anhui Province, Jiangxi Province, Hubei Province, Hunan Province, Chongqing, Sichuan Province, Guizhou Province and Yunnan Province

Table 6.10 Cargo throughput at port and the growth rate of the provinces and municipalities in the Yangtze River Belt for 2009–2013

Province/municipality	2009		2010		2011		2012		2013	
	Throughput (million tons)	Growth rate (%)	Throughput (million tons)	Growth rate (%)	Throughput (million tons)	Growth rate (%)	Throughput (million tons)	Growth rate (%)	Throughput (million tons)	Growth rate (%)
Shanghai	592.0	1.8	653.0	10.4	728.0	11.4	736.0	1.1	775.7	5.5
Jiangsu	1330.0	14.2	1810.0	13.7	2125.9	12.8	1950.0	8.2	2140.0	9.7
Zhejiang	1030.0	9.7	1120.0	8.0	1220.0	8.9	1320.0	7.8	1380.0	4.6
Anhui	260.0	−3.3	330.0	22.9	370.0	15.1	360.0	−3.8	400.0	9.8
Jiangxi	150.1	25.5	211.3	40.8	235.6	11.5	252.7	7.3	262.4	3.8
Hubei	166.7	4.3	187.8	12.7	216.6	15.3	235.2	8.6	262.2	11.5
Hunan	169.4	5.0	142.4	10.0	155.9	9.5	197.3	3.1	210.1	6.5
Chongqing	86.1	9.1	96.7	12.3	116.1	20.0	125.0	7.7	136.8	9.4
Sichuan	51.0	3.7	63.9	25.3	70.8	9.0	77.1	8.9	82.0	6.4
Guizhou	7.8	3.2	8.9	13.8	10.9	23.0	11.6	6.1	9.8	−15.0
Yunnan	3.4	7.6	4.2	2.3	4.6	9.9	3.9	−15.4	49.5	26.9
Yangtze River Economic Belt	3846.5	7.3	4628.2	20.3	5254.4	13.5	5268.8	0.3	5708.5	7.5

Source: Compiled from the *Statistical Bulletin of National Economic and Social Development* (2009–*2013*) of Shanghai, Jiangsu Province, Zhejiang Province, Anhui Province and Chongqing, the *Statistical Yearbook* (*2014*) of Hubei Province, published by the National Bureau of Statistics of Shanghai, Jiangsu Province, Zhejiang Province, Anhui Province, Hubei Province and Chongqing, Statistical Bulletin (2010–*2013*) of Transportation Economic of Hunan Province, published by transportation department of Hunan Province, China Ports Yearbook (*2011–2013*), published by China Port Association

throughput has ranked No. 3 in the world for several consecutive years. In 2014, the airport cargo throughput in the Belt reached 6.24 million tons, rising by 8.7 % year-on-year. Table 6.11 shows the airport cargo throughput and the growth rate of provinces and municipalities in the Yangtze River Economic Belt for 2010–2014.

6.2.2 Transportation Infrastructures

6.2.2.1 Highway Construction

The Yangtze River Economic Belt has formed a well-developed highway transportation network centered in Shanghai, Nanjing, Hangzhou, Hefei, Wuhan, Nanchang, Changsha, Chongqing, Chengdu, Guiyang and Kunming, with expressways as the backbone and spread across the national and provincial highways and county-township roads. By the end of 2013, the highway mileage in operation had reached 1.89 million km, including the 32,000 km national expressways. The target of highway construction for the Belt is to add another 2 million operating highway mileage by the end of 2020 (people.com 2014).

During the 12th Five-year Plan period (2011–2015), the Chinese Government continued to push forward construction of expressways in the Belt. Additional 200 km expressways were newly built in Shanghai and 1,000 km highways were newly built in Chongqing. Other provinces and municipalities in the Belt also invested heftily in expanding their highway networks. The Belt will gradually establish the national expressway network with Shanghai-Chengdu, Shanghai-Chongqing, Shanghai-Kunming and Hangzhou-Ruili expressways, and the common national highway network covering all counties (people.com 2014).

6.2.2.2 Railway Construction

At present, the railway backbone in the Yangtze River Economic Belt is centered in Shanghai, Nanjing, Hangzhou, Hefei, Wuhan, Nanchang, Changsha, Chongqing, Chengdu, Guiyang and Kunming. The main trunk lines of the south-north route include the Beijing-Guangzhou, Beijing-Shanghai, Beijing-Kowloon, and Chengdu-Kunming railways. The main trunk lines of the east-west route include the Shanghai-Kunming, Shanghai-Hangzhou, Hangzhou-Zhuzhou, Zhuzhou-Guiyang and Guiyang-Kunming railways. By the end of 2013, the operating railway mileage in the Belt (including all railways stretching from east to west, or from south to north) reached 29.6 k-km, including the 4,000 km high-speed railway mileage (National Railway Administration of the People's Republic of China 2014).

According to the *Guiding Opinions on Development of the Yangtze River Economic Belt,* China will continue to push forward construction of efficient railway network in the Belt. The country will build the high-speed railway along the Yangtze River connecting Shanghai to Chengdu via Nanjing, Hefei, Wuhan, and

Table 6.11 Airport cargo throughput and its growth rate of provinces and municipalities in the Yangtze River Economic Belt for 2010–2014

	2010		2011		2012		2013		2014	
Province/municipality	Airport cargo throughput (1000 tons)	Growth rate (%)	Airport cargo throughput (1000 tons)	Growth rate (%)	Airport cargo throughput (1000 tons)	Growth rate (%)	Airport cargo throughput (1000 tons)	Growth rate (%)	Airport cargo throughput (1000 tons)	Growth rate (%)
Shanghai	3708.5	24.3	3539.3	–4.6	3368.0	–4.8	3363.6	–0.1	3613.8	7.4
Jiangsu	309.3	18.7	337.8	9.2	368.0	8.9	396.7	7.8	465.1	17.2
Zhejiang	343.3	23.1	365.4	6.4	396.1	8.4	439.3	10.9	481.0	9.5
Anhui	33.4	14.8	40.1	20.1	44.9	12.0	42.9	–4.5	49.5	15.4
Jiangxi	35.5	35.5	38.5	8.5	44.3	15.1	46.6	5.2	54.8	17.6
Hubei	115.2	7.7	129.0	12.0	135.7	5.2	137.5	1.3	152.9	11.2
Hunan	110.2	24.1	117.1	6.3	113.0	–3.5	120.1	6.3	126.7	5.5
Chongqing	195.7	5.2	237.6	21.4	268.7	13.1	280.2	4.3	302.4	7.9
Sichuan	450.6	16.0	496.1	10.1	525.6	5.9	518.5	–1.4	566.9	9.3
Guizhou	61.8	19.8	69.3	12.1	79.7	15.0	77.5	–2.8	83.0	7.1
Yunnan	291.3	7.1	289.7	–0.5	282.6	–2.5	314.6	11.3	340.3	8.2
Yangtze River Economic Belt	5654.8	21.1	5659.9	0.1	5626.6	–0.6	5737.5	2.0	6236.4	8.7

Source: Compiled from the *Statistical Bulletin of Nationwide Airports* (2010–2014), published by the Civil Aviation Administration of China

Chongqing, and the high-speed railway connecting Shanghai and Kunming via Hangzhou, Nanchang, Changsha and Guiyang. In this way, the high-speed railways and express railways in South China and North China will be connected, forming the express railway network which covers all cities with a population of more than 500,000 residents. The country will also re-build and expand the large-capacity regular railways along the Yangtze River, plan the Quzhou-Lijiang Railway and improve the existing capacity of the Shanghai-Kunming Railway. In this way, it would form the passenger-cargo railways which cover cities with a population of more than 200,000 people. It will construct the express railway network with railways along the Yangtze River and with Shanghai-Kunming High-speed Railway as its backbone; it will also establish the regular railway network with railways along the Yangtze River and the Quzhou-Lijiang Railway and Shanghai-Kunming Railway as the backbone. By 2020, the operating railway mileage in the Belt will reach 40,000 km, including 9,000 km of high-speed railways.

6.2.2.3 Waterways

The Yangtze River Economic Belt has many ports that are developing rapidly; it has established a tiered, layered port system spearheaded by Shanghai and centered in major ports in Zhejiang, Nanjing, Wuhan, Chongqing and Jiangsu. In 2014, the Zhoushan Port in Ningbo, Shanghai Port and Suzhou Port ranked No. 1, No. 2 and No. 7 in the world in terms of port cargo throughput. Meanwhile, Shanghai Port ranked first among the world's ports in terms of container throughput volume.

The mainstream of the Yangtze River flows from Western China to Eastern China, with its tributaries connecting the North and South China; they link the hub cities and many small and medium cities and towns in the upstream, mid-stream and downstream reaches of the region, and provide them with easy access via the Yangtze River. The hinterland of water transportation along the Yangtze River includes the seven provinces and two municipalities in the Belt, covering 1.48 million km^2, accounting for 15 % of the total land area in China. In recent years, development of waterway channels on the Yangtze River have been accelerated, with the navigation structure upgraded and optimized, thus promoting coordinated economic growth in the Belt. By the end of 2013, the inland river navigable mileage in the Belt reached 89,000 km, including 6,700 km of top-graded navigation channels. In 2014, the cargo throughput on the mainstream of Yangtze River reached 2.1 billion tons.

6.2.2.4 Airport Construction

The Yangtze River Economic Belt has several large airports such as Shanghai Pudong, Shanghai Hongqiao, Chengdu Shuangliu, Kunming Changshui, Chongqing Jiangbei and Hangzhou Xiaoshan, as well as medium and small airports such as Changsha Huanghua, Guiyang Longdongpu, Huangshan Tunxi and Nanjing Lukou. The Belt has gradually established an aviation network with Shanghai as the international aviation hub and Chongqing, Chengdu, Kunming, Guiyang, Changsha, Wuhan, Nanjing and Hangzhou as the regional aviation hubs. According to the *Plan of Comprehensive Vertical Transportation Corridor in the Yangtze River Economic Belt (2014–2020)*, China will speed up the development of Shanghai international aviation hub while enhancing the role of Chongqing, Chengdu, Kunming, Guiyang, Changsha, Wuhan, Nanjing and Hangzhou as the regional aviation hubs, and ensuring Nanchang, Hefei, Ningbo, Wenzhou, Wuxi, Lijiang and Xishuangbanna play the role as the trunk airports in the region. The *Plan* will also improve the layout of branch airports to form airport clusters in the upstream, mid-stream and downstream reaches of the Yangtze River.

Nowadays, the downstream of Yangtze River has several large-scale cargo airports. For instance, Shanghai Pudong Airport has basically established its position as the international aviation cargo transport hub, with cargo throughput rising most rapidly in Asia Pacific, reaching 2.93 million tons in 2013; it has been ranked No. 1 in China and No. 3 in the world for years. The downstream region also has 21 all-cargo aviation companies and 31 airlines that have cargo flights or chartered cargo flights reaching Pudong Airport. There are 150 sites reachable from Pudong by international (and "regional") cargo flights, and 70 % of the air cargo is transported by all-cargo flights. The proportion of international (and "regional") cargo is 88 %; 58 % of the international (and "regional") cargo in China goes through Shanghai Pudong Airport. The imported and exported cargos mainly include high-tech electronics, autos, textiles and medicines, with a total worth over 120 billion dollars.

Thanks to the strong support from the Central Government and local governments, the mid-stream and upstream reaches along the Yangtze River are also speeding up the construction of airports. For instance, Sichuan Province has invested heavily in airport construction by initiating the preliminary work of Chengdu new airport, which is a national-level international aviation hub. In 2014, the project of building branch airports was approved; construction of the Leshan and Bazhong airports will commence by 2015. During the 13th Five-year Plan period, the Ganzi and Langzhong airports will begin construction.

6.2.3 Development of Logistics Information Platform

Since the 12th Five-year Plan period, logistic informatization has been developing rapidly in provinces and municipalities in the Yangtze River Economic Belt. New information technologies such as big data, cloud computing, Internet, mobile Internet, intelligent logistics have been used extensively, resulting in major changes to the logistic industry. For instance, in 2014, China Mobile Shanghai and Yuan Cheng Logistics Co., Ltd. (one of the first 5A "comprehensive service logistic enterprises" headquartered in Shanghai) formed partnership. The two parties plan to invest 2 billion RMB in the succeeding five years to build the intelligent logistic system, consisting of the innovative logistics management platform, the upgraded supply chain service platform, the industrial Internet platform, and the new-type e-commerce trading platform. China Mobile Shanghai will leverage its technical advantages in IOT, mobile Internet and cloud storage to provide all-around support in technologies, maintenance, service and infrastructures, so as to help Yuan Cheng become a world-class logistic enterprise.

KXTX, a leading enterprise in highway transportation platform, has built its first intelligent logistic park—the KXTX Wuhan Intelligent Logistic Hub. Through IOT technologies and advanced technological equipment, the park will achieve intelligent connections among personnel, vehicles and cargoes, upgrade and optimize the single platform model that focuses on integrated trading in the traditional "highway port" type of logistic park. By so doing, it will build a brand-new "highway port" intelligent operating model that integrates the trunk line resources, improves the branch line delivery, and highlights the Internet-based operations. KXTX has taken the lead in steering the intelligent development of "highway port" logistic parks.

During the 12th Five-year Plan period (2011–2015), the informatization development of navigation system in the Belt has achieved great progress by initiating the construction of intranets for Yangtze River Administration of Navigational Affairs of the Ministry of Transport and the bureaus of maritime affairs, navigation channels, public security, telecom, and the Three Gorges, and establishing the business APP system and management information database. R&D and application projects completed include the "Yangtze River Downstream Navigation Channel Automatic Nautical Chart Mapping and Formation System," the "Yangtze River Three Gorges Lock Escort Channel River Bed Information System," and the "Waterway Passenger Ticketing System." The ports in Nanjing, Nantong, Anqing, Huangshi, Wuhan and Yichang have also built the IT-based system for office work, production, and management. Chongqing Port built the computer network with the intranet, for the central area in Chaotianmen Port and covering seven production and operations companies in the port. In this way, it has successfully developed 17 application systems encompassing the transportation plan and tallying/dispatching operations, thus realizing the network-based development of personnel, finance, production management and office management in the port.

Furthermore, the *Plan of Comprehensive Vertical Transportation Corridor for the Yangtze River Economic Belt (2014–2020)* dictates "taking the lead to build a comprehensive vertical transportation corridor featuring network-based, standardized and intelligent development." The *Plan* also requires establishing cross-regional, cross-departmental and cross-transport-means information sharing, business coordination and service integration, on the basis of improving current management and production service of various transport means in different regions; it also advocated building the inter-connected intelligent logistic network and integrated passenger information comprehensive service platform.

6.2.4 Construction of Logistics Parks

With relatively developed economy, the Yangtze River Economic Belt has several large-scale and outstanding logistic parks, such as Shanghai Pudong Airport Logistic Park, Shanghai Waigaoqiao Bonded Logistic Park, Shanghai Northwest Comprehensive Logistic Park, Yangshan Deepwater Port Logistic Park, Suzhou Chuanhua Logistic Base, and GLP Suzhou Logistic Park. The Pudong Airport Logistic Park is one of the larger and comprehensive modern logistic parks fostered and developed in the downstream reaches of Yangtze River Delta. It consists of logistic warehouses and public service facilities. Its major functional positioning includes international transit, international distribution and delivery, international procurement and international transit trade, while engaging in relevant added-value processing business. As a logistic park with air express, it takes advantage of Shanghai Pudong International Airport's convenient and speedy aviation transport to provide the comprehensive service. The Park is of great significance in enhancing the urban functions and influence on the Belt, and in promoting the development of aviation logistics industry to better serve the whole country.

By 2013, the mid-stream areas in the Belt had 139 logistic parks, ranking No. 1 in China. Of these areas, Hunan has made rapid development in planning and building its logistic parks. Changsha issued the *Plan of Developing Modern Logistic Industry in Changsha (2011–2020),* which mandates the city to plan and build five major logistic parks and ten logistic centers. Of these parks, Changsha Airport Logistic Park is one of the largest logistic parks in the city, occupying an area of 200 hectares. Relying on Huanghua Cargo Airport, the Park provides aviation logistics, distribution and processing, warehousing, delivery, cargo agency and information service for airport-bordering industries and products with high added value. The Park is of great significance in expanding the influence scope of logistics in the mid-stream of Yangtze River and promoting the logistic industry in the Belt.

The upstream region in the Belt is a major hub in the Silk Road. The Government has issued a number of policies to promote the development of logistic parks in the Belt. Sichuan Province released the *Plan of Developing Logistic Parks in Sichuan Province*, which stipulates that the Province will endeavor to build logistic parks featuring a layout having one core, four belts and multiple sites. By planning and building logistic parks, and professional logistic parks or distribution centers specializing in agricultural product, medicine, e-commerce express delivery, and railway transportation that serve local economy and meet people's needs, thus establishing the province-wide logistic park network system that connects the Western China, reaches the entire nation, and faces the world.

6.2.5 Development of Logistics Enterprises

With the location advantage of the Yangtze River, the Belt has attracted a large number of renowned domestic and global logistic enterprises; state-owned, private and foreign logistic enterprises rush to the Belt, developing into a diverse layout of logistic enterprises. State-owned logistic enterprises such as SIPG, Shanghai Railway Bureau, Minsheng Shipping, Chongqing Shipping, CNPC Transportation, Chengdu CREC Southwest International Logistics, Port of Lianyungang Group, and ZJMI are expanding continuously. Private logistics enterprises represented by Shanghai CNEX, Shanghai Deppon Logistics, Jiangsu LSLG and SF Express are also serving well in the logistic market in the Belt. Likewise, well-known international logistic enterprises such as Schenker, FedEx, Clarkson, UPS, TNT, DHL and Maersk have shown up in the Belt.

By 2014, Shanghai had had 98 Grade-A logistic enterprises. In Addition, there were 134 Grade-A logistic enterprises in Jiangsu Province, 123 in Zhejiang Province, 16 in Anhui Province, 12 in Jiangxi Province, 168 in Hubei Province, 61 in Hunan Province, 16 in Sichuan Province, nine in Chongqing, seven in Yunnan Province and one in Guizhou Province. Shanghai, Chongqing, Jiangsu and Hubei also had many 5A and 4A-grade logistic enterprises. Furthermore, Shanghai SF Express, Shanghai Deppon Logistics and Hunan Quanzhou Pharmacy Consumer Goods Supply Chain Co., Ltd. were on the List of Top 50 Chinese Logistic Enterprises in 2014.

6.3 Development of Yangtze River Shipping Logistics

With the golden waterways leading the world inland rivers in cargo throughput, the Yangtze River passage is the most significant east-west axis for regional development in China. Since waterway transportation has the apparent advantages of low cost, large freight volume and low energy consumption, it occupies an important strategic standing in the overall layout of regional development. The Yangtze River shipping logistics is the core and integral part of the comprehensive logistics transport passage in the Belt. It is also the important support for the economic growth and industrial clustering in the Belt. In recent years, the scale of Yangtze River shipping kept expanding, its infrastructures and capabilities to serve the development of the Belt continued improving.

6.3.1 Basic Logistics Conditions of Shipping on the Yangtze River

6.3.1.1 Overview of Shipping Logistics Infrastructures on the Yangtze River

Navigation Channels

Yangtze River is the most developed water system in China; it connects Eastern, Central, Western China, and the southern and northern banks of the River. The navigable mileage on its mainstream and tributaries reaches 71,000 km, accounting for

56 % of the total inland river navigable mileage in China. Of this total, there are 3,920 km navigable channels at above Grade III and 3,130 m^2 navigable channels at Grade IV, accounting for 45.4 % and 46.8 % of the respective grade of waterway mileage in China.

The waterway from Shuifu County to the Yangtze River estuary on the mainstream is 2,837 km long, most part of which is channel above Grade III. From Shuifu to Yichang in the upstream is 1,074 km long, which can provide navigation for inland river ships at 500–3,000 tons; 3,000 ton inland ships can navigate the section from Chongqing to Yichang. In the mid-stream reaches, from Yichang to Wuhan, the 624-km long channel can accommodate navigation for fleets of 1,000–5,000 tons. From Wuhan to Hukou, the 276-km long channel can allow navigation for 5,000 ton maritime ships. In the downstream, from Hukou to Nanjing, the 432-km long channel can provide navigation for 5,000–10,000 ton maritime ships. From Nanjing to Yangtze River estuary, the 432 km portion can provide navigation for 30,000–50,000 ton maritime ships (Qi Tianle 2014). In addition, the Yangtze River navigation channels also include the important regional channel network composed of the tributaries, including Jialing River, Wu River, Min River, Jinsha River, Han River and Jianghan Canal (Yang Bangjie et al. 2015).

Ports and Shipping Centers

On the mainstream of the Yangtze River, there are five shipping logistic centers, namely, Nanjing, Wuhu, Jiujiang, Yueyang and Yibin. They are linked with the development of Shanghai International Shipping Center, Wuhan Yangtze River Mid-Stream Shipping Center and Chongqing Yangtze River Upstream Shipping Center, and work jointly towards developing professional and scaled ports and improving the ports' collection/dispersion/transport channels. Nanjing Port is the major hub port in China and Grade-I port open to the external world. Wuhu Port is one of the major inland river ports in China and Grade-I port; it is also the No.1 coal exporting port on the Yangtze River. Jiujiang Port is one of the water shipping ports in China which is being expanded to open to the outside; it is the sole foreign-trade port in Jiangxi Province that reaches the sea via the River. Yueyang Port is one of the eight best ports on the Yangtze River and the integral part of the Changsha-Zhuzhou-Xiangtan city cluster. The establishment of the afore-mentioned "3+5" shipping centers not only bolsters the shipping capacity but also improves the service efficiency and assurance level of water transport on the Yangtze River (Qi Tianle 2014).

Ports along the Yangtze River have basically established the overall layout, keyed on major national ports, expanded through important regional ports, which

---

[2]According to regulations in the *Inland River Navigation Standards*, the navigation channels in China are rated as I, II, III, IV, V, VI and VII grades. All these seven grades of channels are classified as graded channels. Channels with navigation standards lower than grade VII are classified as non-graded channels.

radiates throughout the entire range of the River. It has thus formed the fairly well-developed river-sea transit system for containers, iron ores and coals, and the professional transport system for auto roll-on and liquefied products. In 2013, the inland river port throughput capacity of the nine provinces and two municipalities in the Belt reached 3.3 billion tons, of which the mainstream of the River accounted for 50 %. The productive berths in the mainstream ports numbered 4,296, of which 459 berths had capacity above 10,000 tons, and one berth possessed the 1 billion-ton capacity. The ships on the River are becoming more standardized, larger and more professionalized. Though the number of ships has declined, the net loading capacity has increased considerably. In 2013, the eleven provinces and municipalities in the Belt had 120,000 inland river ships and hauled 78.3 million tons of load, decreasing by 4.8 % in number, but increasing by 8.6 % by cargo weight from the respective average of 2005 (Yang Bangjie et al. 2015).

6.3.1.2 Logistics Production of the Yangtze River Shipping Business

In 2014, large-sized ports along the mainstream of Yangtze River finished 2 billion tons of cargo throughputs, growing by 7.2 % year-on-year. Of the total, the foreign-trade throughput reached 260 million tons, rising by 5.2 % year-on-year; the container throughput reached 13 million TEUs, falling by 4.5 % year-on-year. The number of passengers carried was 5 million person-times, down by 18.5 % year-on-year (Liaoning Electronic Port 2015). As for various categories of goods shipped, most of the commodities had grown. Figures of (throughput, growth rate) for various categories of goods are as follows: coal and coal products (460 million tons, 5.2 %); oil and gas and their products (84.1 million tons, 0.3 %); metal ores (410 million tons, 8.3 %); minerals and construction materials (340 million tons, 14.1 %); cement (130 million tons, 3.1 %); non-metal ores (80.8 million tons, 5.7 %); chemical raw materials and products (74.4 million tons, 8.4 %). For iron and steel, the throughput was 97.6 million tons, declining by 11.1 % year-on-year.

As for goods shipped by regions, throughput growth rate was fastest in the mid-stream reaches and that in the downstream reaches had the slowest growth. The cargo throughput in the upstream, mid-stream and downstream reached 180 million tons, 210 million tons and 1.6 billion tons, each rising by 8.7 %, 17.0 % and 5.9 %. The growth rates of Luzhou in the upstream, Yichang, Jingzhou, Huangshi and Jiujiang in the mid-stream, and Tongling, Wuhu and Ma'anshan in the downstream all exceeded 10 %. Proportions of port cargo throughput in the upstream, mid-stream and downstream in the total cargo throughput for the mainstream of Yangtze River were 9.0 %, 10.5 % and 80.5 %, respectively, representing a change by 0.1 percentage points, 0.9 percentage points and −1.0 percentage points from the previous year, respectively. As can be seen, the proportion of the downstream was the largest, yet it was less than that in the previous year.

As for goods shipped by sections of the River, the Three Gorges section had 120.4 million tons of cargo, rising by 12.3 % year-on-year; the Three Gorges Lock section had 109 million tons of cargo (not including the equivalent tonnage of passenger ships), rising by 12.3 % year-on-year. Proportions of cargo throughput going upstream and going downstream via the Three Gorges were 56.4 % and 43.6 %, respectively. The freight volume from downstream to upstream rose by 2 % year-on-year, yet its proportion declined by 5.7 percentage points compared with that of the previous year (Chen Liangchao, 2015).

6.3.1.3 Influence of Yangtze River Shipping on the Economy of the Yangtze River Economic Belt

The Radiating and Stimulating Effects

The central cities of Chengdu, Chongqing, Wuhan, Changsha, Hefei, Nanjing and Shanghai linked by the Yangtze River waterways connect the large and medium-sized cities, administrative towns and neighborhood countryside through the radiating influence, thus creating the industrial clustering and scale effects. The banks along the mainstream channel in the Yangtze River waterway are the growth axis of the Belt. The large and medium-sized inland river ports or water-land transport hubs are growth points of major economic centers in the Belt. The cheap and convenient inland river shipping is the principal means of realizing the internal and external exchanges of the Belt. Extensive connections with the economic hinterlands are achieved via the tributary shipping routes and the ports' collection/dispersion/transport routes. Development of Yangtze River shipping not only promotes the optimization of transportation structure of the Belt, but also has far-reaching influence on the industrial and urban structures in the Belt. As the transportation belt which connects Eastern China, Central China and Western China, the Yangtze River shipping is actually the bond connecting the industrial chain in the Belt. The development of Yangtze River shipping plays an irreplaceable role in improving the relatedness of industries and exploring the complementary advantages among the industries along the Belt (You Shibing et al. 2014).

Effect of Industrial Clustering

Though water transport has limitations in scope, many industries and sectors still would gather around the Yangtze River because of the advantages in transportation cost, leading these industries to have relatively strong economic reliance and mutual influence among them. For instance, the processing/manufacturing industry has the characteristics of large raw material needs, high energy usage, and voluminous water consumption. This implies that the development of this type of industries favors scale economy, which coincides with the development attribute of Yangtze River's water transport (You Shibing et al. 2014). Presently, China's economy faces industrial transfer and upgrading. Development of Yangtze River shipping will facilitate the industrial transfer from the coastal areas in Eastern China having higher costs and lower profits to Central and Western China. This transfer will not only promote the industrial development and adjustment of industrial structure of the Central and Western regions, but also help optimize and upgrade the industries in the coastal regions.

6.3.2 Development Characteristics of Shipping Logistics in the Yangtze River

6.3.2.1 Main Logistic Passage for Areas Along the Yangtze River

Yangtze River is the main passage of logistics service for areas along the River, connecting the three logistics regions in the Yangtze River Delta, Central China and Southwest China. The freight volume and freight turnover completed by water shipping on the Yangtze River account for 20 % and 60 %, respectively, of the total freight volume and freight turnover of the areas along the Yangtze River (CPC. people.com.cn). Inland river shipping, owning to its advantages in long distance, large capacity and low cost, plays a crucial role in commodity transport. The shipping capacity on the Yangtze River is equivalent to that of 40 railways. The cost ratio of water transport, land transport and air transport is 1:3:10. Also due to the advantage in friendlier environmental impact, water shipping has become the first choice of large-quantity transport. Presently, water shipping takes 83 % of the power coal transport needed by the power plants along the River, and 85 % iron ore transport needed by the iron and steel enterprises along the River. The mineral construction materials, chiefly low-cost goods such as yellow sands, are traditional source of cargos for water transport (Dong Hongyu 2014). Certain goods, either extra large in dimension, or heavy, or have safety concerns, are difficult to ship via ground transport. Whereas, shipping via Yangtze River affords unmatched advantage in transporting such special goods.

6.3.2.2 Development of River-Sea Transport Accelerated

In September 2014, the State Council issued the *Few Opinions on Promoting Sound Development of Maritime Shipping Industry* and the *Guiding Opinions about Pushing Forward Yangtze River Economic Belt Basing on the Golden Waterways*. These documents signify that pushing forward development of the Belt and promoting sound development of maritime shipping industry have become two major national strategies in China. These two strategies have an intersecting aim of giving water shipping a high priority. To realize these development goals, the crux is to link the domestic and international markets through integrated river-sea transport, so as to provide dependable transport guarantee for China's own economic transformation and upgrading, and for China's economy to be integrated into the world economy. In October 2014, the Ministry of Transport convened the meetings in Wuhan and Shanghai, respectively, to promote the development of the Golden Waterways and the national maritime shipping. In these meetings, the MOT also advocated the elemental concepts of building China into a power of maritime shipping, clarifying the key tasks of facilitating the Golden Waterways construction, and pushing forward the development of the Belt. Consequently, it is imperative to address conceptually of how to match the construction of Yangtze River Golden Waterways with the development of the maritime industry. This means fostering a positive interaction between Yangtze River navigation and maritime shipping to practically realize the integrated development of river-sea transport. These moves will have great practical significance and a far-reaching historical influence (Tang Guanjun 2014).

In recent years, provinces and municipalities along the Belt have been active to develop river-sea transport. For instance, Zhoushan City made full use of its unique location advantage and port resources to develop the river-sea transport, and became integrated into the development of the Belt. In 2013, it realized port cargo throughput of 314 million tons, in which the Ningbo-Zhoushan Port handled 19 million tons of cargo throughput into the Yangtze River ports. The capacity in commodity storage and transit processing at ports has been raised gradually in recent years. The five ports, i.e., the Majishan Ores, Liuheng Power Coal, Century Pacific, Qiaoshan Xingzhong Petro and Laotangshan Transit, have engaged in linking port with shipping; each year, more than 12 million tons of port cargos are supplied to large enterprises of iron and steel, refinery, and power generation in the Yangtze River zone in the form of river-sea transport. Development of shipping business in the Yangtze River zone has achieved an initial favorable result.

During January-August 2014, Shanghai ports handled 10.8 million TEUs in river-sea transport, in which Yangshan Port finalized 5.1 million TEUs of river-sea transport. After the Phase IV project of Yangshan Deep-water Port is completed, there will be more than 10 million TEUs transited via river-sea transport, amounting to a huge market share. At present, the "Shuttle Bus" which engages in river-sea transport, can attain a 20–25 % annual rate of return from operating its container short-barge, thus showing great potential. According to Shanghai's 12th Five-year Plan, by 2015, 45 % of Shanghai Port's throughput would be via river-sea transport. In 2014, the proportion of Yangshan Port's river-sea transport had reached 47.6 % of its total throughput (China Shengsi Government Portal 2015).

6.3.2.3 Inter-port Collaboration Keeps Enhancing

Collaboration and win-win have become the consensus among ports in the Belt. Ports in the Belt are actively responding to the national strategy of pushing forward the Belt basing on the Golden Waterways; inter-port collaborations keep on enhancing. For instance, the Shanghai International Port Group has implemented the "Yangtze River Strategy," through outputting of management know-how, capital and technology, the Group has realized the win-win development of Shanghai Port and other ports along the Yangtze River. At present, it has basically established the logistic service network covering the container wharf, inland tributary container shipping and ship/cargo agencies. Shanghai International Port Group, in the form of capital expansion, participates in the overall reform of Changsha Port and Wuhan Port. By joining the ownership through capital investment, the Group participates in the reform of Jiujiang Port, and is able to actively guide the development of port logistics (LGMI 2008).

In October 2014, the Nanjing Municipal People's government and Luzhou Municipal People's government signed the framework agreement on strategic cooperation of port logistic development in Nanjing. According to the agreement, the two cities and Nanjing Port and Luzhou Port will cooperate in areas of port navigation routes, port Customs clearance and regional logistics (zgsyb.com 2015). In April 2015, Zhoushan City convened the working meeting with the Yangtze River Administration of Navigational Affairs, the MOT, and the CCS on river-sea transport. At the meeting, major ports along the Yangtze River in Chongqing, Yichang, Wuhan, Nanjing and Taizhou and Zhoushan Port released the *Zhoushan Declaration* of River-sea Transport Ports Alliance (cnshipnet.com 2015). The contents of the *Declaration* include: (1) working concertedly to build the river-sea transport logistic public information platform to facilitate information sharing; (2) ensuring the ports to take advantage of their distinctive features and strength and work together to build the mutually-beneficial and win-win logistic partnership mechanism. The Alliance will also strive to coordinate the coastal and Yangtze River ports and port administrations and operators to: (1) relying on the river-sea transport, push forward advantage complementation; (2) enhance exchange and cooperation in major studies, key cargo flows, trade facilitation and grand Customs clearance, thus making active contributions to promoting development of Yangtze River Economic Belt; (3) take advantage of the strength of river-sea transport and negotiate the initiative and key issues of long and mid-term cooperation; (4) work together to push forward the building of market environment featuring fair and orderly competition in the port logistic industry in the Yangtze River, thus guiding and supporting the sustainable, healthy and rapid development of port logistic industry.

6.3.2.4 Accelerated Planning and Construction

In 2014, the Yangtze River Administration of Navigational Affairs finalized the *Guideline Plan of Yangtze River Navigation Development (2014–2030)*, which set the general targets of Yangtze River navigation development for 2020 and 2030. The *Guideline* stipulated the following tasks in seven aspects: building a smooth and efficient navigation system, improving the navigation and service capability in the hubs, improving the comprehensive hub functions of the ports, enhancing the navigation supporting capability, improving the development of navigation market, upgrading and developing the innovation capacity, and building the green and ecological navigation system. Meanwhile, it proposed the implementation assurance measures in four aspects, namely, improving laws, regulations and standards, reforming mechanism and institutions, providing HR and cultural support, and ensuring support for funding policy. Besides, it also finalized the *Plan of Informatization Development in the Yangtze River Navigation,* thus commencing the compilation of the 13th Five-year Plan for the Yangtze River Administration of Navigational Affairs, and conducting studies on 40 planning projects.

In addition, a large number of projects were completed, reviewed and accepted in 2014. The harnessing project for Jiangjiang River navigation channel was 62 % complete, and the main project is expected to be finished by 2015. Great efforts were also made in pursuing quality and untainted projects, safe construction sites, and demonstration projects of ecological and environmental protection. Supervision placed on projects under construction was 100 %; the rate of government projects passing acceptance review was 100 %. No quality safety accidents above alarm-level took place (Yangtze River Administration of Navigational Affairs 2014).

6.4 Summary

The Chinese Government has made the significant strategic decision to push forward the development of the Yangtze River Economic Belt to effectively expand the domestic demand, promote balanced regional economic growth, and strengthen the national economy. This chapter gives an overview of the economic conditions, the current status of the logistic market, and the navigation logistics of the Yangtze River. From the economic development standpoint, the regions in the Yangtze River Economic Belt have taken the opportunities brought on by the Belt being upgraded to the national strategy, leveraged the location advantage and the auspicious governmental policies, and sped up their economic transformation and upgrading. These actions have led to a rapidly-growing economic aggregate and an optimized industrial structure for the Belt. As for the logistic market, the Yangtze River Economic Belt has witnessed a constant expansion of logistic needs, rapid growth of logistics industry, improved logistics infrastructures, enhanced informatization development, and notable improvement in logistics supply capability and service level. Navigational logistics in the Yangtze River Economic Belt has seen expansion in scale, improvement in infrastructures, and collaborations of government offices and among shipping enterprises in the Belt; river shipping via the Golden Waterways has become the crucial support for the Belt's economic growth and industrial clustering.

References

Che Haigang (2014) Time-space meaning of building Yangtze River Economic Belt. Qstheory.cn. http://www.qstheory.cn/jj/qyjj/201405/t20140512_348027.htm. 12 May 2014
Chen Liangchao (2015) Waterway transport economy in the Yangtze River Mainstream gets into new normal. China Waterway Transport News, 25 Jan 2015 (002)
Dong Hongyu (2014) Relying on Yangtze River navigation and developing modern logistic service to serve the development of Yangtze River Economic Belt. Water Transp Manag 10:15–17

Liaoning Electronic Port (2015) In 2014 large-size ports in the mainstream of Yangtze River finalized 1.99 billion tons of cargo throughput, Liaoning Electronic Port. http://www.lneport.gov.cn/2015/27740.html. 26 Jan 2015
Qi Tianle (2014) Study of Yangtze River navigation development strategy in the perspective of waters economy. Sichuan Academy of Social Sciences, Master Thesis
Tang Guanjun (2014) Construction of Yangtze River golden waterways and development of integrated river-sea transport. J Wuhan Tech Coll Commun 16(4):1–4
Wang Rengui (2014) Birth of Yangtze River economic belt strategy. Hexun.com. http://news.hexun.com/2014-09-10/168340009.html. 10 Sept 2014
Yang Bangjie, Yan Yixin, An Xuehui (2015) Analysis and opinions on counter measures for golden waterways problem in the Yangtze River. China Dev 15(1):1–7
Yanhua Z (2014) Yangtze River Economic Belt Brings New Opportunities for Navigation Logistic Enterprises. Ocean Ship Mon 7:60–61
You Shibing, Ren Jingru, Peng Dongfang (2014) Study of interaction between Yangtze River navigation and waterway economy. Reg Econ Rev 4:64–70

Chapter 7
Development of Railway Logistics in China

Jian-hua Xiao

As the artery of national economy, a nation's important infrastructure, and the resource-efficient and environmentally-friendly means of transport, railway is the integral part and the pillar in modern logistic system in China. In recent years, environment for the development of railway logistics in China keeps improving, infrastructures of railway logistics are being bettered, and logistic products are making constant innovations; thus China is gradually establishing a modern railway logistic service system.

This chapter includes four sections. Section 7.1 gives an overview of railway logistics, including its principal characteristics, operating network, main operators and categories. Section 7.2 describes the environment for railway logistics development, including market demands, planning policies, and the relevant standards. Section 7.3 discusses the current development of railway logistics, including market size, infrastructures, channel construction, and information platform. Section 7.4 points out the existing problems and future development trend of railway logistics in China.

7.1 Overview of Railway Logistics

Railway logistics refers to the physical flow process of transporting goods from the origin of supply to the reception destination, mainly on trains by relying on the infrastructures of railway network and stocking yards. Railway logistics encompasses functions of storage, loading/unloading, hauling, packaging, circulation processing, distribution, and information processing.

J.-h. Xiao (✉)
Logistics Research Center, Nankai University, Tianjin, People's Republic of China
e-mail: jhxiao2008@163.com

L. Wang et al. (eds.), *Contemporary Logistics in China*, Current Chinese Economic Report Series, DOI 10.1007/978-981-10-1052-1_7

7.1.1 Main Characteristics of Railway Logistics

Railway logistics is featured by large transport capacity and long transport distance. Specifically, railway transport mainly uses the train carriages hauled by high-powered locomotives, with each train transporting 3,000–5,000 tons of goods. The maximum weight hauled by rail locomotives can reach more than 10,000 tons, much higher than that of air transport or highway transport. As a result, railways can fulfill the task of large-capacity transport. Meanwhile, the average transport distance for Chinese railway freight stands at about 800 km, 25 times more than that of highway transport and 1.15 times more than that of pipeline transport.

Railway logistics features low energy consumption and little pollution. According to the Netherlands Ministry of Transport, the fuel consumption of railway transport is 1.7 liters/100 tons-km, which is far lower than the 4.1 liters consumed in highway transport. The CO_2 emission of railway transport is 48.1 liters/100 tons-km, much lower than the 164 liters consumed in highway transport. Railway transport also features great accuracy and continuity; as a process it is less constrained by natural conditions, thus having great continuity from year-round normal operation. In addition, railway transport also features relatively fixed departure and arrival time, thus having great accuracy. Due to the advantage in large transport capacity and long transport distance, railway freight mainly transports large-volume commodities, such as coal, agricultural products, iron, and steel, which account for 70 % of the total volume of railway freight in China.

Railway transport also has the elements of public-interest and commercialism. The public-interest element implies the social functions railway takes on, such as public transport, macro-controlled transport, and welfare transport. The commercialism aspect means that the railway operator has to take its financial benefits into consideration while providing the service.

7.1.2 Operations Network of Railway Logistics

The general process of railway logistics takes the railway terminal center, transit hub and distribution center as the pivotal nodes, employing transport means such as direct freight trains, express freight trains to connect railways to waterways and highways, thus transporting the freight from production locations to the consumption locations. The railway logistics activities mainly include transport, storage, handling, and distribution of railway freight goods. Figure 7.1 illustrates the operating network of railway logistics.

The express freight train refers to the freight train operated akin to the model of passenger trains, which has fixed departure station and arrival station, fixed train number and operation route, and definite operating cycle and schedule (Hu Yadong 2013). According to the classification of speed, the express freight trains consist of special-express freight train, express freight train and ordinary-express freight train.

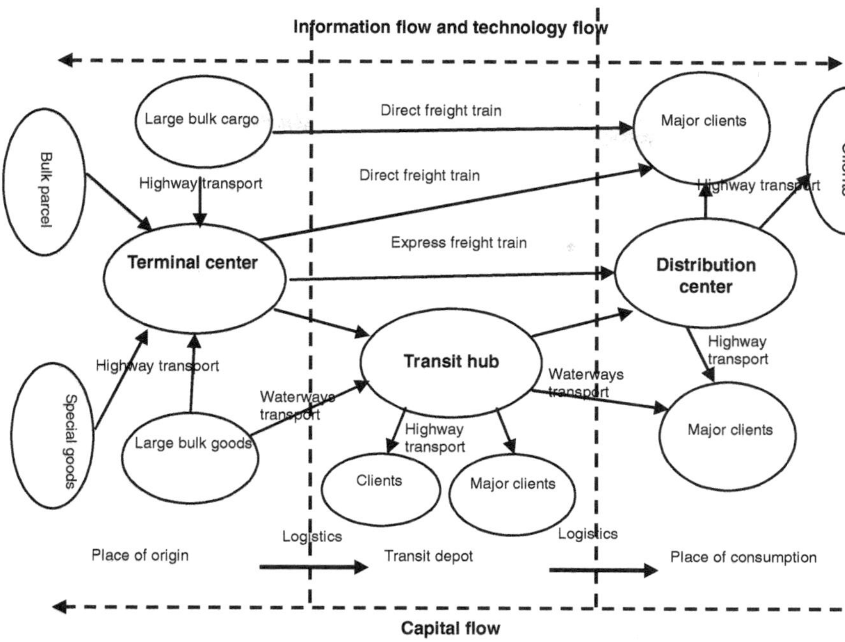

Fig. 7.1 Operating network of railway logistics

The direct freight train refers to the entire array of freight boxcars with fixed "departure station, arrival station, commodity type, cycle, route and train number." The direct freight trains mainly transport bulk cargos featuring large volume, having stable and even source, and whose transport needs can be confirmed in advance, such as coal, coke, petroleum, and metallic ore.

Besides, for clients with bulk parcels, such goods can be transported from the railway terminal center or distribution center via other means of transport such as highways. For clients with large-volume bulk goods, such goods can be transported via the special railway route or other transport means.

7.1.3 Major Operators of Railway Logistics

7.1.3.1 Railway Enterprises

There are three types of railway enterprises.

(1) Enterprises of arterial railway transport that have access to the right of railway freight and railway resources. Such enterprises mainly engage in business of arterial railway transport, e.g., the 18 local railway bureaus subordinate to China Railway Inc., and China Railway Special Cargo Services Co., Ltd. which

mainly specialize in transporting particular goods, and China Railway Container Co., Ltd. which mainly engages in transporting railway containers.

(2) Enterprises which deal in express railway transport of luggage and parcels. These enterprises, such as China Railway Express Services Co., Ltd., engage in domestic inter-city express delivery of luggage, parcels and mails for individual customers.

(3) Comprehensive railway logistic enterprises. Enterprises which have railway background, operating with the third-party logistics operations model, provide holistic and integrated logistics services for large retailers and manufacturers. Such enterprises include China Railway Modern Logistics Technology Co., Ltd., and China Railway United Logistics Co., Ltd.

7.1.3.2 Railway Freight Depots

Railway freight depots are venues that conduct the business of receiving, loading, unloading, storing and pick-up of transported goods. They are places where railway transport is connected to other means of transport. As modern logistic industry develops rapidly, railway freight depots are incorporating the management and service philosophies of modern logistics and supply chains, and actively developing towards railway logistic centers.

7.1.4 Classifications of Railway Logistics

According to the *Classification and Code Table of Railway Freight,* railway logistics includes railway bulk logistics, railway disparate goods logistics, and railway special logistics. Railway bulk logistics includes transport of stable bulk goods such as coal, petroleum, coke, metallic ore, iron, steel, nonferrous metal, non-metallic ore, phosphate ore, mineral building materials, fertilizers and pesticides. Since the stable bulk goods feature stable and even orders, large in transport volume, and having advance confirmation of transport needs, they are usually transported under prior agreements with reserved transport capacity. The disparate goods railway logistics includes transport of goods such as cement, grain, cotton, electronics, electric machinery, fresh and live goods, agricultural products, beverages, tobacco products and medicines. Most of the disparate goods railway logistics are freights with high added value that have high requirements for timely, safe and convenient transport. Its principal service is for extended logistic business and high-end freight goods.

The transport of hazardous goods, the over-sized and over-weight goods, and fresh and live goods are called special railway logistics. Special logistics need to use special vehicles and require special technologies in transporting, loading/unloading, safe-keeping, monitoring, and organizational capability, plus safety and protective measures. In addition, the special railway logistics must also meet the transport, storage, loading, unloading and other requirements of various specific goods.

7.2 Environment for China's Railway Logistic Development

In recent years, China's railway freight market has witnessed a continual trend of declining demand for bulk goods transport, and rising demand for parcel goods and goods with high added value. At the same time, as railway logistics development plans and policies are made and publicized, and as the logistic standards are being improved, the environment for railway logistics development has been bettered.

7.2.1 Declined Market of Bulk-goods Related Industries

Compared with other transport means, railway transport features fast speed, large volume, low cost, and excellent adaptability. For a long period, railway transport has borne the responsibility for the flow and allocation of strategic resources vital to the national economy and people's livelihood. As such, railway freight mainly transports bulk goods such as coal, grain, smelting and agricultural materials. In 2014, volume of coal, iron and steel, mineral ore, and petroleum transport by railway accounted for 60.2 %, 5.4 %, 10.3 % and 3.6 % of the total railway freight volume, respectively, and collectively accounted for 79.5 % of the total railway freight volume (Xiong et al. 2012).

Recently, Chinese economy is shifting from the high-speed growth period to the moderate-speed growth period; growth rate of fixed-assets investment is slowing down, and the growth of added value in major industries such as coal, metallurgy, power, and chemical industry is receding. The railway transport market of "black-goods," typified by coal, has been weak for some time; its volume of railway freight shows a downward trend (Wang Yao 2013). In January-April, 2015, the railway freight in China totaled 0.87 billion tons, falling by 9.4 % year-on-year. The volume of railway freight for 2015 is expected to continue to decline. But in general, the basic demand for transporting bulk goods such as coal, petroleum and smelting materials will not suffer fundamental change. Grain, fertilizer, pesticide, and cotton are materials providing guaranteed demand for rail transport. As a result, during the 13th Five-year Plan period, transport of bulk goods will continue to be the principal market of railway freight.

7.2.2 Robust Market for Disparate"White Goods"

In recent years, while market for bulk goods in China has been weak, market for e-commerce, cold chain, express delivery, food, medicines, home appliances, electronics and autos has been growing rapidly. For instance, in 2014, the revenue of express delivery in China reached 204 billion RMB, growing by 42 % year-on-year. Also in the same year, 13.96 billion pieces were handled in the Chinese express delivery business, gaining 51.9 % over the previous year; for the first time this volume attained the No. 1 ranking, overtaking that of the US. In 2015, the high-speed express delivery trains move among 65 cities in China, transporting 2,000 tons of parcels per day.

The citizens' consumption structure in China has been upgraded in recent years. This will definitely stimulate rapid and continual growth in demand for the transport of general consumer goods. Besides, as technology-intensive industry develops and as processing operations account for the majority portion in the industrial structure, more high valued-added products will be produced, whose proportion in the freight volume will rise. As a result, there will be apparent changes in the market demand for railway freight; demand for transport of bulk goods will continue to fall, whereas demand for transport of disparate "white goods" and high value-added goods will continue to grow.

Furthermore, the market demand for disparate white goods in China are also shifting from the stage of "few categories, large quantity, few batches and long cycle" to the stage of "many categories, small quantity, many batches and short cycle." Such change raises higher requirement on the punctuality, flexibility and reliability of transport services, prompting railway enterprises to make innovations in manners of railway freight production and service, and improve their service level.

7.2.3 Government's Policies in Promoting Market-oriented Railway Development

The 12th Five-year Plan period was very significant for China to deepen its reform and opening-up strategy and accelerate the transformation of economic growth pattern. It was also the important period for China to ensure that the railway industry playing the bedrock role, establishing the comprehensive transport system and supporting the low-carbon economy. In recent years, the Chinese government has issued many policies to facilitate the market-oriented reforms in railways.

In April 2013, the State Council ordained the Sixth Institutional Reform. It established the China Railway Inc. to take charge for unified command of railway transport, conducting railway passenger and freight transport businesses, and taking responsibilities for running dedicated transport and special transport.

In August 2013, the State Council issued the *Opinion about Reforming Railway Investment and Financing Mechanism and Accelerating Railway Development*, which clearly spelled out the following actions: (1) push forward reforms in railway investment and financing mechanism and raise development fund via various means and channels; (2) continue to improve railway transport pricing mechanism and to gradually sort out pricing incoherence; (3) activate railway land resources and encourage comprehensive land development and usage.

In April 2014, the National Development and Reform Commission (NDRC) announced that freight pricing on Zhunchi Railway[1] would follow the market adjustment mechanism, allowing railway transport enterprises, clients and investors to determine the transport price via negotiation. This was the first time for China to liberate the railway transport price, which would play a positive role in reforming railway investment and financing mechanism, and in encouraging and guiding social capital to invest in railway construction.

In December 2014, the NDRC issued the *Notification about Lifting the Control over Prices of Part of Railway Transport Products,* which allowed the market to adjust the price of railway parcel transport and railway disparate goods express delivery. At the same time, the *Notification* also permitted the market to adjust the price of transport on newly-built railways mainly invested by social capitals. The railway transport enterprises can, considering the market demand-supply, production and operating cost, competition, and the society's bearable limit, autonomously determine the price of transport.

In January 2015, the NDRC issued the *Notification of Adjusting Railway Freight Transport and Improving the Mechanism of Forming Prices*, which required: (1) the uniform price of railway freight in China be increased by 0.01 RMB /ton-km; (2) allow the railway transport enterprises to, within the set floating range, determine the specific transport price according to the market demand-supply condition; (3) abolish the collection of "bulk-goods comprehensive logistic service fee."

7.2.4 China Railway Inc. Deepens Its Freight Reform

To ensure railways continue to play the crucial role in the comprehensive transport system, China Railway Inc. has issued a number of policies so as to enhance its reform of the freight institution and deepen the railway freight reform.

In April 2013, China Railway Inc. issued the *Opinion about Pushing forward Reform of Freight Institution*, so as to promote railway freight to move from the "planned model" to the "market model." Specifically, the *Opinion* called for the following actions. (1) Reform the ways of accepting freight orders, simplify the procedures, so that service is provided as orders are placed; (2) Reform the ways of organizing the transport and compile transport program based on the clients' transport needs; (3) Regulate the freight charges; (4) Actively develop the door-to-door full-course railway logistic service and offer the "one-inclusive price" scheme.

In June 2013, China Railway Inc. issued the *Notification of Conducting Door-to-door Railway Freight and Setting and Adjusting Relevant Chargeable Items and Rates*, which clearly dictated expanding the scope of railway transport service and regulating the management of freight charges.

[1] Zhunchi Railway lies in central Inner Mongolia and northern Shanxi Province.

Table 7.1 Railway logistics standards made and issued in 1991–2014 in China

Year	Leading Office	Name of Railway Logistic Standards	Status
1991	Ministry of Railway	*Regulations on Railway Freight Transport*	Issued
1995	Ministry of Railway	*Administrative Measures for Railway Hazardous Freight Transport*	Issued
1999	Ministry of Railway	*Standards of Railway Freight Service Quality*	Issued
2007	Ministry of Railway	*Regulations for Over-sized and Overweight Railway Freight*	Issued
		Regulations for Railway Container Transport	
2008	Ministry of Railway	*Regulations for Loading and Fastening Railway Goods*	Issued
2009	Ministry of Railway	*Regulations for Railway Transport of Fresh and Live Goods*	Issued
2013	China Railway Inc.	*Regulations on Railway Freight Transport (Revised Version)*	Issued
2014	China Railway Inc.	*Provisional Rules on Management of Railway Hazardous Freight*	Issued

7.2.5 *Railway Logistic Standards Are Successively Issued*

To further regulate the railway logistic development environment, and promote standardization of railway logistics, relevant departments and industries of China have actively formulated various railway logistic standards.

In 1991, the Ministry of Railway issued the *Regulations on Railway Freight Transport*, which was the first standard on railway logistics service in China. In the ensuing two decades, the Ministry of Railway issued a number of industrial standards including the *Standards of Railway Freight Service Quality*, the *Administrative Measures for Railway Hazardous Freight Transport*, the *Regulations for Over-sized and Overweight Railway Freight*, the *Regulations for Railway Container Transport*, the *Regulations for Railway Transport of Fresh and Live Goods* and the *Regulations for Loading and Fastening Railway Goods*.

Recently, to meet the needs of reforming the railway freight institution, China Railway Inc., in 2013, revised the inadequate rules and regulations in the *Regulations on Railway Freight*. In 2014, China Railway Inc. issued the new *Provisional Rules on Management of Railway Hazardous Freight*. Table 7.1 lists the railway logistics standards instituted during 1991–2014 in China.

7.3 Current Development of Railway Logistics in China

In recent years, as market environment and policy environment have been continually improved, railway logistics in China have also grown rapidly. Development of logistic infrastructures has been accelerated; to date, China has constructed the grand railway logistic channel consisting of five vertical lines and three horizontal lines. At the same time, railway transport enterprises have developed swiftly, railway logistic service arena has been expanded, and the informatization level of the industry has been upgraded steadily. However, impacted by China's macroeconomic conditions, the scale of railway logistics operations has slid downward in recent years.

7.3.1 Railway Freight Volume Exhibits a Downturn

Because of the recent slowdown of China's economic growth and the nation's economic restructuring, railway freight volume of bulk goods, typified by coal, iron and steel, and iron ore, has declined. In 2012, the railway freight volume stood at 3.9 billion tons, receding by 0.7 % year-on-year, representing a negative growth for the first time in decades (Ministry of Railways of China 2013). In 2014, railway freight volume for China dropped again to 3.81 billion tons, shrinking by 3.9 % year-on-year. Of this total, the volume attributed to national railways was 3.07 billion tons, falling by 4.7 % year-on year (Ministry of Transport 2014). Figure 7.2 shows the volume and growth rate of China's railway freight for 2009–2014.

During the 12th Five-year Plan period (2011–2015), the growth rate of railway freight turnover diminished considerably. In 2012, the railway freight turnover decreased for the first time in decades, standing at 2.92 trillion tons-km, falling by 0.9 % year-on-year. In 2014, the railway freight turnover recorded a decrease again; the total stood at 2.75 trillion tons-km, dropping by 5.6 % year-on-year. Of this total,

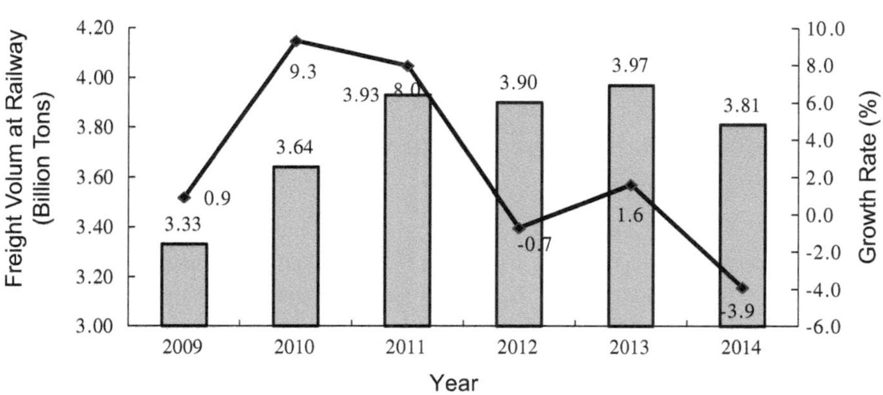

Fig. 7.2 Volume and growth rate of China's railway freight for 2009–2014
Source: Compiled from the *China Statistical Yearbook (2009–2013)*, and the China Statistical Bulletin of National Economic and Social Development (2014), both published by the National Bureau of Statistics of China

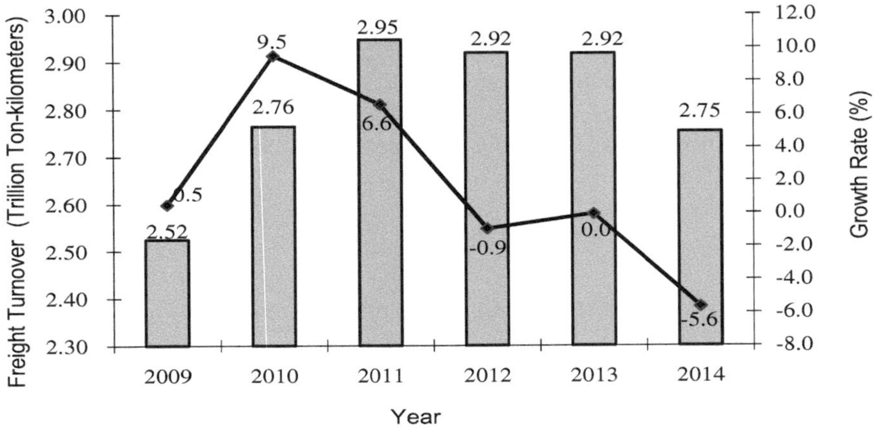

Fig. 7.3 Turnover and growth rate of China's railway freight for 2009–2014
Source: Compiled from the *China Statistical Yearbook (2009–2013)*, and the China Statistical Bulletin of National Economic and Social Development (2014), both published by the National Bureau of Statistics of China

the national railways accounted for 2.51 trillion tons-km, falling by 6.5 % year-on-year. Figure 7.3 shows the turnover and growth rate of China's railway freight for 2009–2014.

Besides, of the total freight market, the proportion of railway freight market kept on declining while the proportions of highway and waterway freight markets were rising continually. Specifically, the market share of railway freight fell from 20.4 % in 1980 to 8.7 % in 2014, sliding by 11.7 percentage points. The market shares of highway and waterway freight have been rising relatively fast. During the same period of time, the market share of highway freight rose from 69.9 % to 76.1 %, and that of waterway freight rose from 7.8 % to 13.6 %.

7.3.2 Railway Logistic Infrastructure Development Being Accelerated

During the past twenty years, the fixed-assets investment in China's railways grew steadily, the railway network continued to expand, and the railway network structure was improved. In 2014, the railway operating mileage in China reached 112,000 km, growing by 8.4 % year-on-year. The proportion of multiple-track railroads reached 50.8 %, up by 4.0 percentage points; the electrification rate amounted to 58.3 %, increasing by 4.2 percentage points year-on-year. The railway network density reached 11.648 km/1000 km^2, rising by 0.904 km/1000 km^2 (Ministry of Transport 2014).

During the same period, China has been accelerating the development of railway logistic infrastructures. A large number of railway container logistic centers and railway logistic parks have been built. For instance, in 2008, the Ministry of Railway issued the *Long and Mid-term Plan of Railway Networks (Revised in 2008),* which stipulated that China would build 18 railway container central stations, 33 secondary container handling stations, and over 100 tertiary container handling depots. By the end of 2014, more than half of railway container centers have been opened and put into operation.

In 2011, the Ministry of Railway issued the *12th Five-year Plan of Railway Logistics,* which explicitly regulated that China would build a triple-level network of railway logistics nodes, namely, the national, regional and local levels. The network would encompass 80 primary railway comprehensive logistic centers, 160 secondary railway comprehensive logistic centers, and 300 professional railway logistic centers. Table 7.2 shows the completed or under-construction railway logistic parks in some regions.

7.3.3 Railway Logistics Channels Gradually Been Formed

China has a large territory and a vast hinterland; its various regions differ greatly in natural conditions and population density. It also has an imbalanced layout of energy, resource and industry. These factors dictate that the long-distance, large-volume and low-cost transport means are imperative to connect the production sites and the consumption markets. Presently, bulk goods such as coal and grain are transported by railways from the producing origins to other distant locations. Currently China has formed a set of railway logistic channels comprised of five vertical lines and three horizontal lines. Figure 7.4 shows the major railway logistics channels in China. Table 7.3 gives more details about these channels.

For a long period, coal transport in China has been relying on railways; coal accounts for about 60 % of all goods transported by railways, ranking atop all other goods. Due to the geographical imbalance between coal production and consumption regions, China has established the railway logistic channels for coal featuring the practice of "Transporting Coal from West to East," and "Transporting Coal from North to South," while utilizing the railway/marine joint transport. The "West to East" railway channels mainly take care of transporting coal produced in Shanxi, Shaanxi, Inner Mongolia, and Ningxia to coal transit ports in North China. The "West to East" railway channels are an integral parts of the "railway/marine transport," which consist of the northern channel, the middle channel, and the southern channel. See more details in Table 7.4.

The "North to South" railway channels are composed of several railways including Beijing-Shanghai railway, Beijing-Kowloon railway, Beijing-Guangzhou Railway, Datong of Shanxi – Zhanjiang of Guangdong railway, Beijing-Baotou railway, Baotou-Liuzhou of Guangxi railway, and Lanzhou-Kunming railway. The Beijing-Kowloon railway is mainly responsible for transporting coal to South China along Shuozhou-Huanghua railway, Shijiazhuang-Dezhou of Shandong railway, Xinxiang of Henan – Rizhao of Shandong railway. The Beijing-Guangzhou railway

Table 7.2 Completed or under-construction railway logistic parks in some regions

Project	Year-Month	Total investment (Billion RMB)	Size in design
Jinan Pingyuan Railway Logistic Base	2013-05 (Construction commenced)	0.2 (Phase I)	It includes six functional zones – coal, bulk cargo, container, parcel, bonded, and chemical zones. The designed logistic capacity for coal logistic functional zone is 9 million tons/year.
Jining Railway Logistic Park	2013-12 (Project approved)	3.94	It covers about 2 km^2 with 0.6 km^2 of building area. Planned buildings are railway logistic zone, shipping logistic zone, management service zone, and highway transport and distribution center.
Suining West Railway Logistic Park	2013-03 (Construction commenced)	3	It covers 1 km^2, including the railway logistic zone, logistic warehousing and highway/railway transport zone, customs supervising zone, logistic trading zone, cold-chain logistic zone, comprehensive service zone and other functional zones. The designed cargo throughput is 8 million tons/year. In 2015, the cargo throughput will reach 5 million tons/year.
Heze Railway Logistic Park	2015-01 (Construction commenced)	1.97 (Phase I)	The project covers 3 km^2 with 2 km^2 for Phase I and 1 km^2 for Phase II. The Phase I project has ten dedicated lines and Phase II project has11 lines reserved.
Urumchi Railway International Logistic Park	2014-04 (Construction commenced)	5	The total building area is 120,000 m^2, including the merchandise display shopping mall, commercial logistic distribution center, business and financial service center, logistic information center, and life supporting zone.
Guangdong Qingyuan Railway Logistic Park	2015-02 (Project approved)	3	The project covers about 1.3 km^2, which includes railway logistic zone, logistic warehousing and highway/railway transport zone, cold-chain logistic zone, comprehensive service zone, and life supporting zone, etc.

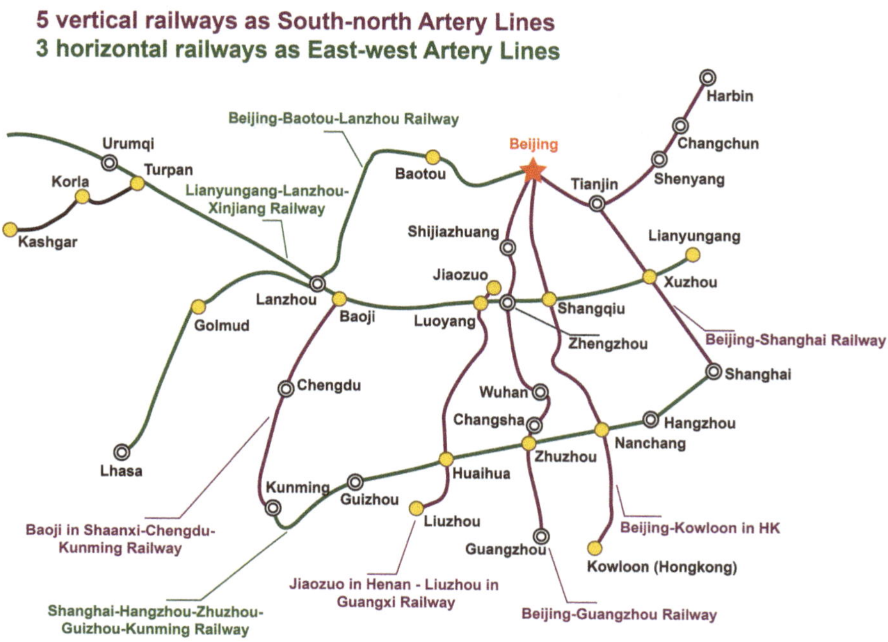

Fig. 7.4 Map of five *vertical* and three *horizontal* railways in China

mainly transports coal to South China along Shijiangzhuang-Taiyuan railway and Lanzhou-Lianyungang railway in Jiangsu. The Datong of Shanxi-Zhanjiang of Guangdong railway lies in the transition part between Central China and Western China. It is the important channel that transports coal from Shanxi, Shaanxi and west of Inner Mongolia to Central China and Southern China. (Wu Yunliang and Huang Shaopeng 2008).

Besides, China is accelerating the development of international railway channels. It has formed the international railway logistic channels based on Alataw Pass, Erenhot, Manchuria and other ports of entry. See Table 7.5 for Chinese border railway ports of entry.

7.3.4 *More Innovative Railway Logistic Products*

During the 12th Five-year Plan period, China Railway Inc., in considering the characteristics of systematic and intensive transport needs for bulk goods, proposed ways of organizing railway transport based on the punctuality system. Also by attending to the "multiple-item, small-quantity and multiple-batch" demand attributes of goods with high added value, it proposed ways of operating freight trains

Table 7.3 Major railway logistics channels in China

Freight channel	Composition of channel	Goods category
Five vertical lines	Beijing-Guangzhou Railway	South-bound cargos mainly include coal, iron and steel, timber and exported goods; north-bound cargos mainly include rice, nonferrous metal and imported goods.
	Beijing-Kowloon in HK	Cargos mainly include mineral resources, grain, cotton, oil and agricultural products produced in regions along the railway.
	Jiaozuo in Henan – Liuzhou in Guangxi Railway	Regions along the railway produce grain, cotton, oil, tobacco and other agricultural products, as well as minerals such as coal and nonferrous metal. The railway plays an important role in helping improve railway layout, upgrade the ability of transporting coal out of Shanxi, and divert the transport volume of Beijing-Guangzhou Railway.
	Beijing-Shanghai Railway	South-bound cargos mainly include coal, iron and steel, timber and cotton; north-bound cargos mainly include machinery, meters and general merchandise, etc.
	Baoji in Shaanxi-Chengdu-Kunming Railway	The cargos mainly include minerals and timber produced in regions along the railway.
Three horizontal lines	Beijing-Baotou-Lanzhou Railway	West-bound cargos mainly include iron and steel, machinery and timber; east-bound cargos mainly include coal, ore and livestock products.
	Lianyungang-Lanzhou-Xinjiang Railway	Cargos mainly include coal, petroleum and other minerals produced in regions along the railway, as well as cotton and livestock products.
	Shanghai-Hangzhou-Zhuzhou-Guizhou-Kunming Railway	East-bound cargos mainly include grain, timber and nonferrous metal; west-bound cargos mainly include iron and steel, machinery, cement and general merchandise.

Table 7.4 "Transporting coal from the West to the East" railway logistic channels

Coal transport channel	Composition of channels and their tasks
Northern channel	It consists of Datong of Shanxi-Zhungeer Qi of Inner Mongolia railway, Beijing-Baotou railway, Datong of Shanxi-Qinhuangdao of Hebei railway, Baotou-Shenmu of Shaanxi railway, Shuozhou of Shanxi-Huanghua of Hebei railway and others. These railways mainly transport power-generating coal from the west of Inner Mongolia, north of Shaanxi and north of Shanxi. It is the main channel of transporting coal out of Shanxi, Shaanxi and west of Inner Mongolia.
Middle channel	It consists of Beijing-Yuanping of Shanxi railway, Shijiazhuang-Taiyuan railway, Taiyuan-Jiaozuo of Henan railway and Handan of Hebei – Changzhi of Shanxi railway, etc. The channel mainly transports coke and anthracite from the middle part of southwest of Shanxi.
Southern channel	It consists of Taiyuan-Jiaozuo railway, Houma of Shanxi-Yueshan of Henan railway, Lanzhou-Lianyungang railway, Xinfeng Town of Shaanxi-Xunyang Town of Shanxi railway and Yongning of Nanjing-Xinfeng Town of Shaanxi railway, etc. The channel mainly transports coke, fat coal and anthracite from the middle part and southwest of Shanxi.

Table 7.5 China's border railway ports of entry

Country	Port of entry	Location	Approved to open in	Comments
Mongolia	Erenhot Port of Entry	Erenhot City, Inner Mongolia	1956	Grade-1 national port of entry; the sole railway port of entry between China and Mongolia; the important strategic hub on the Euro-Asia Continental Bridge.
Russia	Manchuria Port of Entry	Manchuria City, Inner Mongolia Autonomous Region	1901	The railway port of entry in China with the largest scale and highest through capacity.
	Suifenhe Port of Entry	Suifenhe City, Heilongjiang	1903	As the grade-1 national port of entry, it takes the task of China-Russia joint international transport.
	Hunchun Port of Entry	Hunchun City, Jilin	2003	Grade-1 national port of entry.
Kazakhstan	Alataw Pass Port of Entry	Bortala Mongol Autonomous Prefecture, Xinjiang	1990	Grade-1 national port of entry, international joint transport port of entry, and west bridgehead on the Chinese section of the new Euro-Asia Continental Bridge.
	Khorgos Port of Entry	Kazak Autonomous Prefecture of Yili, Xinjiang	2012	China's International railway channel opening to the West.
DPRK	Ji'an Port of Entry	Ji'an City, Jilin Province	1954	Important railway port of entry connecting China and DPRK.
	Tumen Port of Entry	Tumen City, Jilin Province	1954	
	Dandong Port of Entry	Dandong City, Liaoning Province	1955	
Vietnam	Hekou Port of Entry	Honghe Prefecture, Yunnan Province	1952	The convenient railway channel connecting China and ASEAN; the bridgehead connecting Euro-Asia Continent and ASEAN railway continental bridge.
	Pingxiang Port of Entry	Guangxi Autonomous Region	1952	

in the fashion similar to passenger trains. Such operations format includes the following five forms.

(1) Express delivery of disparate goods. In September 2014, China Railway Inc. launched the regional disparate goods express-delivery trains, which basically covered the entire country. More than 70 regional cargo express-delivery trains were set off in the Northeast China region, the Beijing-Tianjin-Hebei region, the Yangtze River Delta, and the Circum-Xinjiang region. On the artery railways such as Beijing-Shanghai railway, Beijing-Guangzhou railway, Beijing-Harbin railway and Shanghai-Kunming railway, 16 cross-regional freight express-delivery trains were put into operation. Between the neighboring Railway Bureaus in other regions, many branch lines of freight express-delivery trains were put into operations. By February 2015, two million units of disparate goods weighing more than 70,000 tons were regularly transported on railways per day, and a total of 4.34 million tons of such goods had been transported on railways.
(2) E-commerce trains. The railway departments formed partnership with e-commerce enterprises and express delivery enterprises to operate six e-commerce trains, at maximum speed of 160 km/h, between Beijing, Shanghai, Guangzhou and Shenzhen, and transport 500,000 small parcels each day.
(3) China-Europe international trains. China Railway Inc. initiated the China-Europe international channels from Western, Central and Eastern China. The Western Channel starts from Central and Western China and leaves China via Alataw Pass (Khorgos). The Middle Channel passes North China and leaves China via Erenhot. The Eastern Channel passes the coastal areas in Southeast China and leaves China via Manchuria (Suifenhe). In March 2011, the Chongqing-Duisburg (Germany) scheduled train, the first China-Europe international train, began operation. Later, Chengdu, Zhengzhou, Wuhan, Suzhou, Yiwu and other cities also started the China-Europe scheduled trains. In 2014, China Railway Inc. dispatched 308 China-Europe trains, a number escalated by 285 % year-on-year. In 2014, China Railway Inc. delivered 26,070-TEU containers.
(4) High-speed railway luggage and parcel transport. It mainly provides door-to-door transport service for documents, and small and light express-delivery goods less than 5 kg in weight.
(5) Total-process logistics. In recent years, China Railway Inc. has actively developed the door-to-door total-process railway logistic service. First, it built the door-to-door collection and delivery network; second, it adopted the one-inclusive price system for the total-process door-to-door logistic service, so as to usher the railway freight towards modern logistics.

7.3.5 Railway Transport Enterprises Continually Expanded

During the 12th Five-year Plan period, China Railway Inc. concertedly supported the enterprises to adopt various measures such as regrouping, merging, capital augmentation, and share expansion to drive the traditional railway transport enterprises

Table 7.6 List of grade-5A logistic enterprises in China

Grade	Enterprise	Headquarters
5A	China Railway Special Cargo Services Co., Ltd.	Beijing
5A	China Railway Container Transport Co., Ltd.	Beijing
5A	China Railway Express Co., Ltd.	Beijing
5A	China Railway Modern Logistics Technology Co., Ltd.	Beijing
5A	Shanghai Railway Bureau	Shanghai
5A	Beijing Railway Bureau	Beijing
5A	Taiyuan Railway Bureau	Taiyuan
5A	Shenyang Railway Bureau	Shenyang
5A	Jinan Railway Bureau	Jinan
5A	Zhengzhou Railway Bureau	Zhengzhou
5A	Wuhan Railway Bureau	Wuhan
5A	Nanning Railway Bureau	Nanning
5A	Chengdu Railway Bureau	Chengdu
5A	Kunming Railway Bureau	Kunming
5A	Xi'an Railway Bureau	Xi'an
5A	Urumqi Railway Bureau	Urumqi
5A	Lanzhou Railway Bureau	Lanzhou

to transform into modern logistic enterprises. China Railway Baggage and Parcel Express Delivery Co., Ltd., China Railway Special Cargo Services Co., Ltd., and China Railway Container Transport Co., Ltd. had grown rapidly into logistic enterprises with great clout. For instance, in 2014, China Railway Container Transport Co., Ltd. adopted the strategy of "Going Abroad" to boost the brand name of China-Europe trains. For 2014, it transported 26,078 TEUs totaling 150,000 tons of goods worth 1.85 billion USD.

By February 2015, China Railway Inc. had possessed 17 grade-5A logistic enterprises, as shown in Table 7.6. Among them, China Railway Container Transport Co., Ltd., China Railway Modern Logistics Technology Co., Ltd., China Railway Special Cargo Services Co., Ltd. and Zhengzhou Railway Economic Development Group Co., Ltd. were on the list of Top 50 Chinese Logistic Enterprises in terms of main business revenues for 2014, ranking No. 21, No. 27, No. 31, No. 34 and No. 37, respectively.

7.3.6 *Informatization of Railway Logistics Moves Forward Steadily*

Informatization development of railway logistics has been carried on continually in recent years. Logistic IT devices such as video capturing, bar code, RFID, and hand-held terminal have been applied in railway freight depots. The Station

Comprehensive Management System, Train Number Identity System, Train Forecasting and Confirmation Report System, Transport Dispatch Information System, and Railway Office Information System have been further upgraded.

In September 2012, China Railway Freight E-commerce Platform was put into nationwide use. This platform provides online handling of railway freight business. Clients, via the system, can inquire about railway freight information, submit transport needs, reserve logistics service, and book railway freight services.

In April 2014, the Kunming Railway Bureau launched China's first Railway Freight WeChat Service Platform to propel freight service into the "Palm Era." The platform, by connecting with railway freight e-commerce system, can help owners of goods to inquire about the status of transported goods, train dispatching, published transport resources, and other freight services; it can also publish information such as the range of service offers and transport fee calculation. (Zhou Liujun 2014)

In May 2015, China Railway 95306 Freight Website was put online. The website mainly provides three services: (1) railway freight e-commerce service, transport fee inquiry, cargo tracking and other railway freight services; (2) bulk goods trading service, and support online trading of coal, ore, iron and steel, grain, chemicals, cement, coke, fertilizer, timber, beverage and food, as well as related auxiliary logistic service; (3) small-commodity trading services, including shopping and purchasing of goods, online payment, logistics distribution, online marketing, and customer service.

7.4 Problems and Trend of Railway Logistics in China

Though railway logistics in China has made evident progress, there are still problems, such as low level of market development, backward depot facilities and equipment, and yet-to-be-improved service functions, that need to be resolved as soon as possible. As the environment for railway logistics development keeps improving, reform of the railway freight market will be sped up. China will pay more attention to and secure better development of railway international logistics, railway logistics information platform, and railway logistics infrastructures.

7.4.1 Problems in China's Railway Logistics

7.4.1.1 Low Level of Market-oriented Railway Logistics Development and Inadequate Service Functions

At present, the marketization level of railway logistics is low, as reflected in the following three aspects. (1) The investment and financing for railway logistics infrastructures are mainly made by the China Railway Inc., the Shenhua Group and other enterprises with SOE background, with limited involvement by the non-governmental capital. (2) Railway logistics enterprises are mostly traditional-type railway transport enterprises; few are large professional third-party railway logistics enterprises. (3) The all-inclusive fee of door-to-door railway transport is still higher than that of highway transport. The railway transport charges need to transform towards marketization.

Besides, railway logistic service functions need improvement, as indicated in the following three aspects. (1) Most railway transport enterprises are deficient in providing modern logistic services featuring comprehensive, total-process and integrated development; (2) Service functions of railway freight depots still manifest pronounced features of traditional basic freight transport functions, lacking diverse value-added logistic services such as warehousing, distribution, circulation, processing, and packaging. (3) The door-to-door total-process business now operated by railway departments integrated the staff from the Diverse Business Department[2] and from the freight transport areas of the Loading and Unloading Department; these staff workers have less know-how in value-added services such as circulation, processing, logistic information service, stock management, and logistic cost control.

7.4.1.2 Backward Depot Facilities and Equipment, Low Level of Informatization

In earlier times, railway depots were designed to ensure speedy arrival and departure of goods to improve the transport efficiency. Nowadays, railway depots are confronted with the following problems. (1) Warehouses and storage yards have insufficient capacity with aged loading and unloading and warehousing facilities; (2) Forklifts, pallets, handsets and other modern logistic equipment are in short supply, resulting in low mechanization level of loading and unloading; (3) Part of the newly-operated depots are only equipped with loading and unloading tracks, cargo platforms, and canopies, yet are lacking relevant loading and unloading machineries, measuring and safety inspection equipment, and information system.

In addition, the level of logistics informatization is in need of enhancement. The IT foundation of railway freight is lagging behind the development of modern logistics. The total-process logistic system fails to provide the complete door-to-door transport information. The internal operations management information systems need to be integrated and improved. Presently there is no railway logistic public information platform oriented to the clients (Xie Xushen 2014).

---

[2] The diverse business means being extended from the primary business in railway industry to other fields in the form of side industries.

7.4.1.3 Disconnection with Other Transport Means; Marine/Railway Joint Transport Needing Improvement

Major technical standards and operating capability of railway depot facilities and equipment do not match or supplement those of other transport means such as highway and waterway. Linkages of loading, unloading, connection, delivery and receiving are not smooth, causing low operations efficiency. Some highways leading to the railway container central stations and roads inside the stations have low hardening standards, thus hindering the "seamless linking" of railway transport and other means of transport (Zhao et al. 2014).

Besides, railway takes up low percentage in port container transport because of the impact caused by railway transport capacity, capability of reloading facilities, and mismatched frontend and backend collection and distribution systems. At present, highways and inner river waterways capture most of the transport work of port containers. Railway transport gets less than 5 % of port container throughput, which is far less than the 30–40 % mark attained in developed countries.

7.4.2 Trend of Chinese Railway Logistic Development

7.4.2.1 Reform of Railway Freight Market Will Make Further Breakthroughs

During the 13th Five-year Plan period, sourcing structure and transport demand of railway freight will be altered gradually. The total demand for bulk goods transport will continue to fall; the "white goods" market will gradually become the new growth point in the railway freight market, posing new requirements for the ways of organizing transport. Consequently, the reform of railway freight organization will continue to deepen, thus making further breakthroughs in railway freight market reform. The development of railway logistic products, investment and financing of logistic infrastructures, and the reform of freight pricing mechanism will speed up the market-oriented development.

Besides, facing China's need for deepening the overall reform, China Railway Inc. will likewise keep deepening the internal management system reform, including clarifying the relationships between China Railway Inc. and the railway bureaus, confirming their respective duties and responsibilities, and regulating the management behaviors of their respective legal entities. In this way, China Railway Inc. can institute the internal management system that complies with the features of railway transport enterprises, optimize the organizational structure, and push forward the transformation of railway transport enterprises towards modern logistic enterprises that better meet the market needs (Zhang Weidong and Wei Xiaojin 2014).

7.4.2.2 International Railway Logistics Will Gain Rapid Growth

Railways serve as the major carrier of land transport in China's "One Belt and One Road" strategy. In recent years, a number of Europe-Asia logistic channels have been opened, including the Chongqing-Xinjiang-Europe, the Wuhan-Xinjiang-Europe, the Zhengzhou-Xinjiang-Europe, the Chengdu-Xinjiang-Europe and the Tianjin-Xinjiang-Europe channels. In the form of continental bridge, railways keep extending into the hinterland of Europe, thus greatly improving the efficiency of international logistics connecting China with Europe and Middle Asian countries. At the same time, because the New Europe-Asia Continental Bridge[3] channel has the advantages of shorter transport time than maritime transport and lower cost than air transport, it has become an important international railway container transport channel.

With the One Belt and One Road strategy moving forward, development of the logistic channel keyed on China-Europe Railway and China-Asia Railway will be accelerated. Railways will gradually gain connectivity, the Customs clearance environment will become more convenient, and more international railway freight trains will be put into operation. Hence, the international railway logistics will attain faster growth.

7.4.2.3 Informatization Development of Railway Logistics Will Be Accelerated

Railway logistics features numerous information sources and great information dynamics. It is essential to build a safe, efficient, full-functioned, modern railway logistic information system with smooth information flow and extensive coverage.

The informatization development of railway logistics must integrate the information systems of all relevant businesses, employ new technologies such as GPS, GIS, GPRS/GSM, RFID and EDI, to achieve supply-chain resource sharing, communal use of information, real-time tracking, effective control, and total-process management of all sections of the railway logistics.

In addition, China is also accelerating the integration of railway logistics with e-commerce. Railway freight e-commerce platform will be improved. In concrete terms, the platform can process e-payment and transfer settlement, provide real-time feedback of business operations and query of goods en route information, complete online delivery formalities, and uniformly handle customer complaints and insurance claims. (Zhao Yi 2013).

---

[3]The New Europe-Asia Continental Bridge refers to the second Europe-Asia Continental Bridge, which is the convenient railway that connects the maritime transport routes on the fringes of Asia and Europe. It starts with Rizhao and Lianyungang in China on the east, and leaves China at Alataw Pass in Xinjiang on the west. Then the railway goes through Kazakhstan, Russia, Belarus, Poland, Germany and reaches Rotterdam Port in the Netherlands. It is now the most convenient channel that connects the west and the east of the Europe-Asia continent.

7.4.2.4 Pace of Railway Logistic Infrastructures Construction Will Pick up

In recent years, the robust development of domestic railway network help free up the transport capacity. In particular, the construction of passenger railways will provide a leeway for the development of freight logistics. According to NDRC, during the 13th Five-year Plan period, China will continue to deepen the reform of railway investment and financing mechanism, continue to push forward large railway projects in Central and Western China, and inter-city railways. The scale of railway network will continue to expand and the structure of railway network will continue to improve.

Furthermore, local railway bureaus, while striving to upgrade and renovate their existing freight depots, will hasten the construction of a number of modern railway logistic centers featuring diverse functions, modern equipment, convenient procedures, and total-process services. Meanwhile, the local railway bureaus will also accelerate the construction of transit facilities and make great efforts to develop multi-modal joint transport, in particular the marine/railway joint transport and the highway/railway joint transport.

7.5 Summary

This chapter first introduces the basic characteristics, operating network, major players and the classifications of railway logistics in China. Next, it elaborates on the development environment of railway logistics, including the changed demand in railway freight market, the relevant policies and plans issued by the government and the China Railway Inc., and the standardization of railway logistics. Then it discusses the current development of railway logistics, which has made evident progress in infrastructure development, logistic channels construction, product innovation, and the level of informatization development. Last, it points out the problems which China's railway logistics currently faces, such as low level of market development, backward depot facilities and equipment, and insufficient service functions. It also proposes some key issues in China's railway logistics future development.

References

Hu Yadong (2013) Push forward and secure better and faster development of container railway/waterway transport. New Silk Road Horiz 1:29–31

Ministry of Railways of China (2013) The railway statistical bulletin (2012). Ministry of Railways of China, Beijing. 13 Mar 2013

Ministry of Transport (2014) The statistical bulletin of transportation industry (2014). Ministry of Railways of China, Beijing. 30 Apr 2015

Wang Yao (2013) Analysis of current situation of railway freight market and counter-measures. Railw Freight 5:11–14

Wu Yunliang, Huang Shaopeng (2008) China's coal logistics network system: optimization and policy side suggestions. China Coal 10:27–33

Xie Xushen (2014) Thinking on how to transform railway freight into modern logistic base. Railw Freight 5:1–6

Xiong Lingling, Gu Hai, Zhang Yi (2012) Research on current market status and development strategy of railway freight transport. Sci Technol Ind 4:1–6

Zhang Xiaodong, Wei Xiaojing (2014) Review of 2013 railway logistic development and outlook in 2014. China Log Procur 11:60–63

Zhao Haikuan, Wang Tao, Song Kai (2014) Speed up logistic-oriented reform of railway depots and improve ability and quality of railway logistics. China Railw 9:1–5

Zhao Yi (2013) Thinking on optimizing and expanding railway freight products. Shanghai Railw Sci Technol 2:7–84

Zhou Liujun (2014) Development status and countermeasures about road LTL logistics in Liuzhou. Value Eng 15:36–37

Chapter 8
Development of Air Logistics in China

Ran Wei

Since the twenty-first century, China has sped up the transformation of its economic growth pattern and the upgrading of its industrial structure. The demand of air logistics market has been growing continually. Air freight enterprises are speeding up their logistics-oriented transformation and air logistics is pushing forward its opening-up to the outside world. Meanwhile, China's market environment of industrial development is gradually improved. With market demand for air logistics continually rising and many government policies being issued to promote its development, air logistics in China has attained rapid growth in recent years.

This chapter has four sections. Section 8.1 briefly states the major characteristics of the operations flow of air logistics. Section 8.2 analyzes the market environment for the development of China's air logistics in recent years. Section 8.3 sums up the current development and existing problems in air logistics. Section 8.4 presents an outlook of the future trend of air logistics in China.

8.1 Major Characteristics and Operations Flow of Air Logistics

As an integral part of modern logistics, air logistics refer to the entire process of transporting raw materials, finished products and relevant information effectively from the starting point to the terminal point, by employing air transport as the main means of transport, and using modern information technologies to connect the supplier and demander parties. Air logistics organically integrates transportation, warehousing, loading/unloading, processing, sorting, distribution, and information to

R. Wei (✉)
College of Economy and Management, Civil Aviation University of China,
Tianjin, People's Republic of China
e-mail: weiran2350@sina.com

L. Wang et al. (eds.), *Contemporary Logistics in China*, Current Chinese
Economic Report Series, DOI 10.1007/978-981-10-1052-1_8

form a complete service supply chain, thus providing the multi-functional and integrated comprehensive service for users. Different from other mainstream means of logistics, air logistics possesses characteristics as discussed in the following sub-sections.

8.1.1 Air logistics Transports Goods Rapidly and Mainly Serves High-Valued Products

Air logistics involves logistic activities relying mostly on air transport and has the obvious feature of transitional speed. For instance, inter-continental express freight often arrives within 1–5 days, and ordinary domestic express freight arrives within 1–3 days. Such delivery speed is not easily attained in other means of logistics. Because air freight capacity is very much limited by the transport vehicle, air logistic cargoes are developing towards smaller size, lighter weight and higher value. Most of the goods transported by air logistics are fresh and live products, precision machineries and electronic products. According to data published by the International Civil Aviation Organization, in 2013, the global air freight industry carried 49.30 million tons of goods, accounting for only 1.0 % of the global trade volume. However, the goods transported by air logistics were worth 6.8 trillion USD, accounting for 33 % of the world trade value (caacnews 2014). Thus, the goods transported by air logistics tend to exhibit the feature of high values.

8.1.2 Air Logistics Highlights Integration of Logistic Functions

Air logistics focuses more on the integration of logistic functions and the coordination of partnerships. Compared to traditional air freight, modern air logistics enterprises not only have to consummate the air transport of goods, but also actively participate in the clients' whole logistics operations process. Besides, they also make overall design and management of clients' logistic operations targeted at reducing costs.

Air logistics is most featured by adopting the value-chain operating model and highlighting the logistic functional integration. Modern air logistic service is targeted at meeting the needs of clients by providing the door-to-door comprehensive freight service predominately through air transport. The various service enterprises that are involved in air transport, goods collection and delivery, door-to-port and port-to-door ground transport, and depot service form a complete value chain.

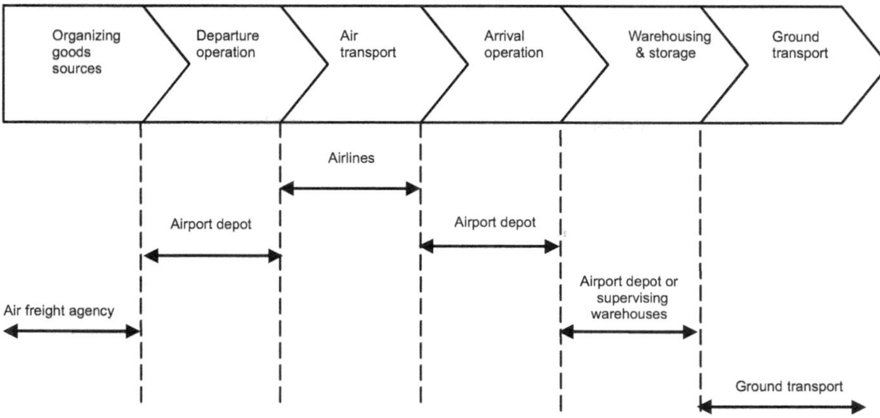

Fig. 8.1 Operations flow of air logistics

8.1.3 Air Logistics Spans Large Space and Has International Characteristics

Because air logistics transports quickly over a long range, and the layout of air routes is not so much constrained by the geography, most of the air logistics transports are long-distance and wide-spanned. As a result, air logistics is predominantly international in nature. Take the air freight turnover in China as an example. In 2013, the civil air freight turnover stood at 17.03 billion tons-km, of which the freight turnover of international routes was 10.92 billion tons-km, accounting for 64.1 % of the total; the average transport distance of air freight is 3,034 km (National Bureau of Statistics of China 2014).

The operations flow of air logistics starts with the organizing of goods sources, followed by such processes as goods departure, air transport, goods arrival, goods storage, and goods delivery. Participants in the whole process include air freight agencies, airport depots, airlines, and ground transport enterprises. This operations flow can effectively realize flow of goods, shift responsibility for safekeeping the goods and exchange the information. Figure 8.1 illustrates the operations flow of air logistics.

8.2 Market Environment of China's Air Logistics Development

Since the twenty-first century, China has sped up the transformation of its pattern of economic growth and the upgrading of its industrial structure, triggering a steady rise in the market demand for air logistics. At the same time, the Government keeps issuing a number of various policies and measures that encourage the development of air logistics, thus pushing forward the opening-up of air logistics. Also, the market environment for the air transport industry has also been gradually improved.

8.2.1 Market Demand for Air Logistics Continues to Rise

The long-term, steady and rapid growth of China's economy has stimulated a continual rise of the demand for air logistics. Since the "reform and opening-up," and up until two years ago (1979–2012), China's economy had maintained a rapid growth at an average growth rate of 9.9 %. The rapid economic growth has brought about the rapid rise in the demand for air freight.

The rapid growth in the foreign trade in China has also stimulated the incessant rise in the demand for air logistics. In 2013, the foreign trade volume stood at 4.16 trillion USD, rising by 7.6 % year-on-year, making China the largest country in trading of goods, and the sole country in the world whose trade volume exceeded 4 trillion USD.

During the past five years, e-commerce and express delivery business have been growing explosively in China, generating the nation's new growth spots for the demand of air logistics. According to the data published by the State Post Bureau, in 2014 China shipped 13.96 billion pieces in express delivery business, rising by 52 % year-on-year, and ranking No. 1 in the world. As e-commerce develops rapidly, goods for e-commerce lead to rising needs for express transport. As a result, express delivery enterprises have become the largest clients of air logistics enterprises. In 1997, express-delivery goods accounted for only 4.5 % of the air transport market. However, this proportion rose to 16.4 % in 2009 and escalated to 40 % in 2014.

8.2.2 "Open Sky" Boosts the Opening-Up of China's Air Logistics

Since the establishment of the "Open Sky" international air transport policy in 1995, the United States has signed numerous bilateral agreements with many countries and regions following the definition of "Open Sky." The U.S. "Open Sky" action has produced far-reaching influence on the air logistics in China, and directly pushed forward the "Opening of Traffic Right" of air logistics in China. Meanwhile, the Chinese Government, by instituting policies and signing international air transport treaties, has promoted the opening-up of its air logistics market as well.

In 2002, the Civil Aviation Administration of China (CAAC) issued the *Regulations on Foreign Business Investing in Civil Aviation Industry,* which stipulated that the share of foreign investment in public air transport enterprises in China shall not exceed 25 %. In 2005, CAAC issued the *Opinion about Certain Policies and Measures to Promote International Aviation Transport,* which specified giving priority to opening up the international air transport market. According to the *Opinion*, (1) newly-established Chinese air freight enterprises, if meeting the standards of safe operations, can directly apply to operate international air routes; (2) initiate the working mechanism of National Air Transport Procedures Simplification Committee, and actively coordinate with the exit and entry administration offices of Customs, border control, and quality inspection; (3) according to relevant standards of the International Civil Aviation Organization, improve the inspection procedure and solve problems pertaining to prompt clearance of goods, shuttle trucks and electronic customs declaration, etc.

In September 1980, the Chinese and the U.S. Governments signed the *China-US Civil Aviation Transport Treaty*. Since then, China had successively signed with the U.S. in 1999, 2004 and 2007, the *New China-US Civil Aviation Transport Treaty*, thus expanding mutual opening of the air transport market. By the end of 2013, China had signed 115 bilateral air transport treaties with other countries and regions.

In terms of Opening of Traffic Right,[1] China has enhanced its flexibility in making air transport arrangements with other countries in recent years. In 2003, with approval by CAAC, Hainan became the pilot-test province in China which opened the 3rd, 4th and 5th Traffic Right. By 2015, China has opened the 5th Traffic Right for air freight in Xiamen, Nanjing, Shanghai, Tianjin, Wuhan, Zhengzhou and Yinchuan, attracting many foreign airlines to open the international air freight routes connecting China to a third country. Besides opening the Traffic Right, CAAC also encourages the Chinese and foreign airlines to establish partnership in the form of flight code sharing and joint venture.

8.2.3 Various Policies and Measures Advancing Air Logistics Development Were Issued

In recent years, the State Council, the Civil Aviation Administration of China and local governments in China have issued a number of policies to encourage air logistics development, which have greatly promoted the development of air logistics industry in China.

In 2012, the State Council issued the *Certain Opinions on Promoting the Development of Civil Aviation*, which specified air logistics development as one of the major tasks in the future civil aviation development. In 2013, the General Office

[1] "Traffic Right" is a concept in international air transport, appeared first on the 1944 International Civil Aviation Conference in Chicago; it is also termed the "Right of the Free Sky." Historically, setting the "Traffic Right" is an instrument for protecting a nation's aviation resource. In today's trend of globalized economy, not opening the "Traffic Right" is construed as a hindrance to a country's air transport. The 3rd Traffic Right refers to the landing right of passengers and cargos in the destination; the 4th Traffic Right refers to the right of passengers and cargos departing from the destination; and the 5th Traffic Right refers to the right as intermediate point and extension node.

of the State Council issued the *Plan for Division of Tasks to Promote Development of Civil Aviation Industry,* which stressed the fact that in order to enhance its international aviation competitiveness, China must research the policies of opening up international air transport, and encourage the strong passenger and freight airlines in China to enhance their international competitiveness.

In 2013, the State Council approved the China (Shanghai) Pilot Free Trade Zone. Of the 98 major tasks listed in the *Master Plan of China (Shanghai) Pilot Free Trade Zone*, nearly 20 reform or pilot-test tasks are related to air transport and airport.

In September 2014, the State Council issued the *Long-term and Mid-term Plan of Logistic Development for 2014–2020,* which clearly regulated "supporting express-delivery industry to integrate resources and make concerted efforts with civil aviation, railway and highway so as to form a number of large express-delivery enterprises with international competitiveness," and "supporting air freight enterprises to make merger and acquisition or re-organization so as to grow stronger and larger, thus improving the comprehensive logistics service capability."

The Civil Aviation Administration of China, as the administrative office of the industry, has been issuing supporting policies to provide strong policy guarantee for China's air logistics enterprises. These policies cover different fields, such as independent right of operations, development of air routes, selection of flight nodes, market guidance, promoting competition, fostering of market, induction of transport capacity, network building, layout of industrialization, and allocation of Traffic Right.

In 2010, CAAC issued the *Strategic Vision of Establishing a Powerful Aviation Country.* The document clearly outlined the target of freight development, e.g., "encouraging freight airlines to make merger and acquisition, re-organization, or business partnership, thus improving the international competitiveness of China's air freight industry; seeking to have at least one freight airlines in China being listed among the Top Five in the world by 2030."

In 2014, CAAC issued the *Opinion of Pushing Forward Coordinated Development of Civil Aviation in Beijing, Tianjin and Hebei* so as to implement China's strategy of coordinated development of the Beijing-Tianjin-Hebei region, and build the rapid, convenient, efficient, safe, large-capacity and low-cost civil aviation integrated system in the region. The document stipulates the improvement of the level of aviation assurance and transport service in the Beijing-Tianjin-Hebei region and pursues the coordinated development of civil aviation and regional economy.

To support a speedy development of the aviation air logistics industry, local governments in China have issued many preferential policies to encourage the development of air logistics. In 2006, Wuhan became the first pilot test region of comprehensive air transport reform in China. According to the *Plan of Implementing Civil Aviation Transport Comprehensive Reform Pilot Test in Wuhan*, Hubei Province and Wuhan City adopted several policies in areas such as airport tax refund and exemption of airport construction fees. They conducted rational planning of land ground transport network in the airport, and worked proactively toward the development of Wuhan Aviation Town, so as to attract airlines and large logistic enterprises to build operating bases in Wuhan. In 2012, Henan Province, to support the rapid development of air logistic industry in the Province, established the Zhengzhou Aviation Economy Comprehensive Pilot Test Zone. While working hard to obtain preferential policies from the Central Government, it also instituted 20 local preferential policies to encourage the development of air logistics.

8.2.4 Civil Aviation Regulations and Industry Standards Are Gradually Been Improved

Since 2012, CAAC, basing on the *Law of the People's Republic of China of Civil Aviation,* the *Regulations of Domestic Transport of Goods via Chinese Civil Aviation* and the *Regulations of International Transport of Goods via Chinese Civil Aviation,* has been improving the regulations, systems and industrial standards in civil aviation transport.

In 2012, CAAC issued three industrial standards, i.e., the *Transport Standards of Works of Arts and Museum Exhibits by Air,* the *Specification of Consolidated Consignment by Air,* and the *Specification for Air Cargo Agents Service.* These three standards regulate the general rules of transporting art works and museum exhibits, and the specifications for packing, labeling, marking, transporting, and handling of aberrant transport; they also stipulate the documentation of air transport goods, the limitations and requirements for the collection and delivery of goods, and the regulations of sales and services of air cargo agents.

In 2013, to fortify the management of transporting hazardous goods by air, promote its development and ensure the safety of air transport, CAAC issued the *Administrative Regulations of Transporting Hazardous Products via Chinese Civil Aviation* (Civil Aviation Administration of China, Decree No. 216). The *Regulations* made explicit specifications on the limitations of transporting hazardous goods by air transport, its approval procedures, and the preparations and responsibilities of the shipper, operator and agent. In 2014, CAAC issued the *Transport Standards of Cold Chain Cargo by Air,* which regulated the classification of temperature range, the equipment and facilities of cold-chain air transport, as well as the requirements for packaging, labeling, operations, shipping documents, goods reception, ground storage, warehouse, apron area ground transport, loading, temperature control during flight, unloading, delivery, handling of aberrant shipments, transport information, and services.

8.3 Current Development of Air Logistics in China

Lately as market demand for air logistics continues to grow and as China keeps on issuing policies that promote air logistics, air logistics has seen rapid growth in China, in spite of the weak global economy. Air logistic infrastructure development

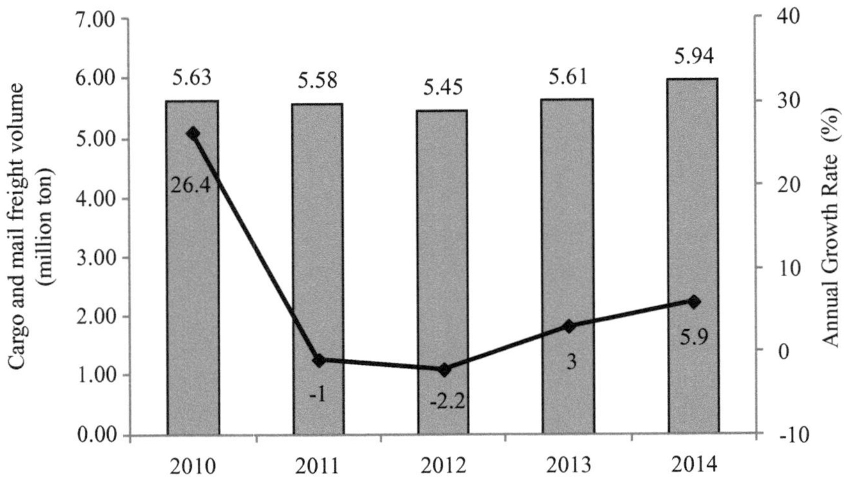

Fig. 8.2 Civil aviation cargo and mail freight volume for 2010–2014
Source: Compiled from the *Statistical Bulletin of Development of Civil Aviation (2014)*, published by the Civil Aviation Administration of China

in China is gaining speed and hub airports across the nation are playing more prominent roles. Air transport enterprises are making faster transformation towards logistics-oriented development. The center of gravity of air logistics is gradually shifting from the coastal areas in the East to the Central and Western regions.

8.3.1 Air Logistics Volume Is Growing Steadily

Due to the weak global economy and the European debt crisis, air freight volume in China declined slightly in 2011 and 2012. Yet in the latter part of the 12th Five-year Plan period (2011–2015), air freight volume rebounded and key civil air transport indicators continued to rise steadily.

According to the *Statistical Bulletin of Development of Civil Aviation* issued by CAAC, in 2014, China attained 5.941 million tons of civil aviation cargo and mail freight volume and 18.78 billion tons-km of cargo and mail turnover, increasing by 5.9 % and 10.3 %, respectively, from the previous year. In 2014, civil airports in China handled 13.56 million tons of cargo and mail throughput, up by 7.8 % year-on-year. Of this amount the domestic air routes accounted for 8.855 million tons, rising by 6.7 % year-on-year (the air routes originating from mainland China to Hong Kong, Macau and Taiwan was 905 k-tons, rising by 16.0 % year-on-year); the international air routes accounted for 4.706 million tons, rising by 9.8 % (The Civil Aviation Administration of China 2014a). Figure 8.2 shows the civil aviation cargo and mail freight volume for 2010–2014.

8.3.2 *Infrastructures Develop Faster and Hub Airports Play a Greater Role*

In the past five years, CAAC and the governments of all levels have hastened the development of air logistic infrastructures, both in number and in size of airports. In 2010, the airport system in China completed 44.15 billion RMB of fixed-assets investment. In 2013, the airport system completed 50.75 billion RMB of fixed-assets investment, rising at an annual average rate of 4.8 % in three years. In 2014, the airport system finalized 56.08 billion RMB of fixed-assets investment, rising by 10.5 % year-on-year (The Civil Aviation Administration of China 2014b). By 2014, there were 202 certified civil aviation airports in mainland China (not including Hong Kong, Macau and Taiwan).

During this period, the airport cargo and mail throughput were gradually concentrated in a few airports. In 2014, there were 50 airports each with an annual cargo and mail throughput over 10,000 tons, and whose total freight throughput accounted for 98.5 % of the total volume of the 200-plus airports in China. Airports in Shanghai, Beijing and Guangzhou are playing a significantly prominent hub role; cargo and mail throughput of these three cities accounted for 51.3 % of the total in China. The freight throughput in Shanghai Pudong International Airport ranks No. 3 in the world. Table 8.1 lists the rankings of cargo and mail throughput for China's civil airports in 2014.

8.3.3 *Air Freight Enterprises Are Accelerating the Transformation to Logistic-Oriented Development*

During the past five years, air freight enterprises in China have launched respective strategies of transforming to logistic-oriented development. Air China Cargo was the first one to work out the comprehensive cargo transport solutions to connecting China to the world. China Cargo Airlines Ltd. was the first airlines that launched the strategy of "United Sky & Earth" in China highlighting taking advantages of its air transport capability, and meanwhile enhancing its ground transit operations in the Yangtze River Delta and high value-added transport business. Moreover, in a creative way, it puts forward the transformation concept of "express delivery + e-commerce + trading." China Southern Airlines Cargo has signed partnership agreement with YTO Express and SF Express and joined SkyTeam Cargo in November 2010. In the beginning of 2011, Hainan Airlines launched the Super X program to integrate its cargo resources via Grand China Logistics. By and large, all major air freight enterprises in China have taken their first step towards logistic-oriented development.

Table 8.1 Rankings of cargo and mail throughput for China's civil airports in 2014

Airport	Ranking	2014(kiloton)	2013(kiloton)	Growth(%)
Shanghai/Pudong	1	3181.7	2928.5	8.6
Beijing/Capital	2	1848.3	1843.7	0.2
Guangzhou	3	1454.0	1309.7	11.0
Shenzhen	4	963.9	913.5	5.5
Chengdu	5	545.0	501.4	8.7
Shanghai/Hongqiao	6	432.2	435.1	−0.7
Hangzhou	7	398.6	368.1	8.3
Zhengzhou	8	370.4	255.7	44.9
Kunming	9	316.7	293.6	7.8
Xiamen	10	306.4	299.5	2.3
Nanjing	11	304.3	255.8	19.0
Chongqing	12	302.3	280.2	7.9
Tianjin	13	233.4	214.4	8.8
Qingdao	14	204.4	186.2	9.8
Xi'an	15	186.4	178.9	4.2
Urumqi	16	162.7	153.3	6.2
Wuhan	17	143.0	129.5	10.5
Shenyang	18	138.3	136.1	1.7
Dalian	19	133.5	132.3	0.9
Changsha	20	125.0	117.6	6.3

Source: Compiled from the *Statistical Bulletin of Nationwide Airports (2014)*, published by the Civil Aviation Administration of China

8.3.4 Industry Migration Out of Eastern China Sways Air Logistics Focus to Central and Western China

Along with the implementation of the "Go West Strategy" and the transformation and upgrading of the industrial structure in the Eastern Region, many manufacturing enterprises have moved from the coastal areas in Eastern China to Central China and Western China. Concurrently, the focus of air logistics has also begun to shift to Central China and Western China.

Since 2011, air cargo has evidenced a trend in which the growth rate of airports in Central China and Western China is faster than that in Eastern China. Take Zhengzhou as an example. After Foxconn settled in Zhengzhou, Zhengzhou Airport achieved 370 k-tons of cargo and mail throughput in 2014, rising by 44.9 % year-on-year; its ranking jumped from No. 15 in 2012 to No. 8 in 2014, leading all airports in growth rate. In March 2015, Chengdou Shuangliu Airport Cargo Airlines Transit Center, built by the Sichuan Airport Group with a total investment of 20 million RMB, was put into use. The Transit Center has attracted a number of heavyweight enterprises such as SF Express to locate there. SF Logistics also decides to build its Western China Air Hub Transit Center in Shuangliu Airport, which will elevate the influence of Chengdu as the air logistic hub in the West.

8.4 Problems and Trends of Air Logistics in China

Though China's air logistics has been developing rapidly in recent years, there are a few problems such as singular operating mode, yet-to-be-improved service functions, scattered business operations, and low IT development, which are barriers to better development of air logistics in China. However, the outlook of air logistics is bright -- market size will continue to expand, air logistics enterprises will speed up the pace of merger and acquisition and re-organization, and IT/IOT technology applications will also make breakthroughs.

8.4.1 Problems in China's Air Logistics Development

8.4.1.1 Air Logistics Operating Mode Is Singular and Service Functions Need Improvement

At present, most airports and cargo service of airlines adopt the indirect operating mode. Collection and delivery of air-transported goods are mostly handled by air cargo agencies. The main business of traditional air cargo companies is to carry out the air transport of goods; these companies don't get involved in the clients' logistics operations process. As a result, they are unable to provide holistic design and management of logistics operations of the clients whose aims for logistics operations are cost reduction. The service functions for these traditional air logistics enterprises are in need of improvement.

8.4.1.2 Operations of Air Logistics Enterprises Are Disparate and International Air Transport Market Share Is Shrinking

For a long time, most of the domestic air cargos and mails were carried in the cargo compartment of passenger planes. Generally speaking, a cargo fleet with properly balanced carrying capacity has not been established; the number of cargo planes is scanty and the cargo planes are scattered. By the end of 2013, the 97 cargo planes in civil aviation market in China are spread out among seven cargo airlines; the top airlines held barely over twenty cargo planes. In 2013, out of all types of air cargo enterprises, five cargo and passenger airlines had all-cargo planes, with a market share of 56 %; thirty-six cargo and passenger airlines which had no all-cargo planes, with a market share of 30 %; three all-cargo airlines, with a market share of 7 %; two express-delivery cargo airlines, with a market share of 7 % (Caacnews 2014). The transport capacity of cargo compartments on passenger planes were scattered over many airlines.

Because the Chinese air cargo operations are scattered and lag behind the overseas air logistics enterprises, they are evidently disadvantaged in the international market competition. In the 11th Five-year Plan period (2006–2010), the Chinese airlines had more than 30 % share in the international (including Hong Kong, Macau and Taiwan) market. However, the market share continued to shrink in the 12th Five-year Plan period (2011–2015). In 2013, this market share was down to 20.6 %.

8.4.1.3 Level of Informatization of Air Logistics Is Low and Public Information Platform Is Insufficient

Independent cargo information systems are available for the airlines, hub airports, and large cargo agents that constitute the supply chain of China's air logistics. However, smaller airlines, airports and most of the cargo agents lack such information system. The existing information systems are not compatible with one another, thus unable to facilitate information sharing on a broader scale, and there is no industry-wide air logistics public information platform. As a result, the air logistics system has fragmented information resources and snagged information flow, making it difficult to fully attain the overall functional advantage. In general, the informatization level of air logistics in China is comparatively low, which is mainly manifested in the narrow scope of informatization application of the industry, sluggish development, low level, and insufficient capital investment.

8.4.2 Trends of Air Logistics Development in China

8.4.2.1 Market Size Continues to Expand and Freight Structure Continues to Lean Towards Higher Value

For years, China has been in the lead in world's air freight market. And it will keep its leading position in the future development. Over several upcoming years, the Chinese market will still be one of the fastest-growing markets in the world, particularly in its express-delivery markets. As the Chinese enterprises become more international, the "Just-in-time Production" and the "Zero Inventory" management models in the global manufacturing industry will exert increasing needs for the global supply chain solutions. This will lead to greater reliance on comprehensive logistic enterprises, and need sizable global air logistics systems to support them.

The optimization and upgrading of industrial structure in China will also stimulate continual growth in demand for transporting higher value-added products, which have higher requirements for the speed and quality of transport. Consequently, the air freight structure will transform towards higher value-added products typified by the electronic products.

8.4.2.2 Logistics-oriented Development of Air Freight Shifting Towards Vertical Integration and Horizontal Coordination

The logistics-oriented development of air freight will be reflected by the following two orientations. (1) Realize integrated logistics by extending the vertical chain; this is achieved by the internal integration within the enterprise so that clients can have access to the one-stop service. (2) Realize multiple-party coordination by subcontracting to the horizontal businesses, known as functional logistics subcontracting. Air logistics operators adopting this approach will sign a general logistics service contract with their clients, and are responsible for organizing the entire business process of logistics. They subcontract the actual functional business of logistics by subcontracting to enterprises engaging in various logistics operations.

8.4.2.3 Air Logistics Enterprises Speeding Up Merger and Acquisition and Reorganization

In 2014, the State Council issued the *Long-term and Mid-term Plan for the Development of Logistics Industry, (2014–2020)*, which clearly stipulated "supporting air freight enterprises to make merger and acquisition or reorganization, so as to grow stronger and bigger, thus improving the comprehensive logistics service capability." Guided by this policy, the air logistic enterprises will enhance cooperation among enterprises in the future, so as to realize resource integration in a holistic way. The more influential air logistic enterprises will also, through M&A and reorganization, become stronger and bigger. In the future, the air logistics market in China will eventually have two or three commanding air logistics enterprises that take the dominant position and capture most of the market share.

8.4.2.4 IT and IOT Technology Applications Will Make Breakthroughs

Technological development and applications can propel leap-forward development of the industry. As a key element in the new-generation of technological revolution, the IOT technology can not only gradually end the "information-silo" phenomenon in China's air logistics, but also achieve its integration into modern business model. As IOT technology matures, air IOT technology will make breakthrough applications in the air logistics field. In 2014, Hainan Airlines went into partnership with Tsinghua National Laboratory of Information Science and Technology to build the National Air Transport IOT Application Demonstration Project, employing the IOT technology to improve the service quality of air transport. The demonstration project encompasses the five airports of Hainan Airlines and all air routes of Hainan Airlines, Tianjin Airlines and Capital Airlines.

8.5 Summary

In recent years, air logistics in China has witnessed a mode of leap-forward development. Total volume of air logistics has been growing steadily, air logistic infrastructures are speeding up the construction, hub airports are playing a greater role, and air cargo enterprises are facilitating logistics-oriented development. The focus of air logistics is gradually shifting from the coastal areas in Eastern China to Central and Western China. Amid its rapid development of air logistics, China also confronts problems of singular operating mode, imperfect service functions, scattered business operations, and low level of informatization, which seriously impede the healthy development of its air logistics industry. With on-going rapid development in air logistics, it is expected that the scale of air logistics market will continue to expand, air logistics enterprises will speed up the pace of merger and acquisition and reorganization, and the application of IT and IOT technologies will make notable breakthroughs.

References

Caacnews (2014) Summary of research report on development of Chinese aviation logistics, CAAC news. http://www.caacnews.com.cn/newsshow.aspx?idnews=261864. 22 Dec 2014

National Bureau of Statistics of China (2014) China statistics yearbook (2014). National Bureau of Statistics of China, Beijing

The Civil Aviation Administration of China (2014a) The statistical bulletin of nationwide airports (2014), the Civil Aviation Administration of China, Beijing. http://www.caac.gov.cn/I1/K3/201504/t20150403_73469.html

The Civil Aviation Administration of China (2014b) The statistical bulletin of development of civil aviation (2014), the Civil Aviation Administration of China, Beijing

Chapter 9
Development of China's Pilot Free Trade Zones

Jing-lei Yang

Free Trade Zone is the platform for a country or region to open up to the outside world and an important carrier for it to get involved in the international competition. It is also the important node in the global supply chain network. Pilot Free Trade Zone development in China focuses on institutional innovation and explores new experience for the deepening of reform and expansion of opening-up on a full scale. Till today, China has achieved notable results for the initial phase in transformation of government functions, trade supervision, foreign investment management, financial innovation, and legal system construction. Institutional reform and innovation not only became the fundamental force boosting logistics development in Pilot Free Trade Zone, but also instilled new dynamics for China to uplift the competitiveness of its logistics enterprises, speed up its regional logistics integration and explore the high-end logistics services.

This chapter discusses the development of Pilot Free Trade Zones in China in four sections. Section 9.1 introduces the macro-economic background of Pilot Free Trade Zone development and the history of Pilot Free Trade Zones in China. Section 9.2 elaborates on the setup, operations and major construction results of China (Shanghai) Pilot Free Trade Zone (Shanghai FTZ). Section 9.3 presents the highlights and characteristics of the development plans of Pilot Free Trade Zones in Guangdong, Tianjin and Fujian. The last section discusses the current status and future development of logistics in the Pilot Free Trade Zones.

J.-l. Yang (✉)
Business School, Nankai University, Tianjin, People's Republic of China
e-mail: jingleiyang@163.com

L. Wang et al. (eds.), *Contemporary Logistics in China*, Current Chinese Economic Report Series, DOI 10.1007/978-981-10-1052-1_9

9.1 Overview of Pilot Free Trade Zones in China

9.1.1 Macro-economic Environment for Establishing Pilot Free Trade Zones in China

9.1.1.1 Domestic and Overseas Economic and Trade Background for Establishing Pilot Free Trade Zones

Presently, the world economy is still going through the process of unstable recovery. Recovery in the US is strong yet growth in the European Union and Japan is slowing down; growth in developing economies is not all optimistic either. However, the mild recovery trend of world economy has basically been established and the trend of economic globalization has remained unchanged. Because of the slow recovery of world economy and economic globalization, the world trade volume has been growing at a low speed. Developed countries have become the major support to global trade and international investment. Meanwhile, the import needs of developing countries are declining, and the geopolitical crises in some regions are also impeding the trade recovery in certain regions (World Economic Outlook 2014). After experiencing rapid economic growth for over two decades in the past, China has been in the stage of economic restructuring for the last two years. As a result, the traditional dynamism for China's economic growth has been weakening and the foreign trade is in a severe decline. The export products, mainly of processing trade, are in the low end of the industry chain in the international market, thus having moderate competitiveness, and the service trade is apparently lagging behind (China's Foreign Trade Situation 2014).

9.1.1.2 New Changes in the International Trade Rules

Facing the new global economic development environment, the developed nations are actively seeking new trade advantages. In terms of WTO's trade system construction, the Special Meeting of WTO General Council passed, in November 2014, the Trade Facilitation Agreement, which was the first Global Facilitation Agreement reached by WTO since its establishment more than two decades ago (Xinhuanet 2014). The Agreement proposed increasing the cross-border trade efficiency and lowering the operating cost of global supply chain, so as to promote global trade development and build the "Single Window" mechanism to simplify Customs and port of entry clearance procedures.

About building bilateral and regional trade systems, developed economies such as the US, EU and Japan have pushed forward Trans-Pacific Partnership Agreement (TPP), Transatlantic Trade and Investment Partnership (TTIP), Trade in Service Agreement (TISA), Pluri-lateral Services Agreement (PSA), and other regional trade agreements. These agreements encompass many fields including tariff, trade, investment, service, labor, environment, information, and non-tariff barriers, and have important contents on market access and regulation, regulatory consistency, Technical Barriers to Trade (TBT), and sanitary and phytosanitary measures (Li Chunding and Zhao Meiying 2015).

The frequent changes and adjustments of international trade rules, on the one hand, have made global trade more smooth and convenient. On the other hand, developed nations have also curbed the role of emerging countries in the international trade negotiations. The building of Pilot Free Trade Zones in China is an important measure for actively participating in the development of multi-lateral trade system and catering to the new situation of global trade system development.

9.1.2 Development Process of the Pilot Free Trade Zone in China

9.1.2.1 Initial Stage of Reform and Opening-Up

In the beginning period of reform and opening-up, China established the Special Economic Zones and Economic Development Zones. By reducing taxes for certain enterprises, exempting income taxes for certain period of years and offering other preferential policies in the above Zones, China aimed to attract investment, create new jobs and boost its economic growth. To cater to the needs for entrepot trade and simple processing for foreign trade, these Zones have bonded warehouses to keep bonded goods.

As foreign trade develops rapidly, the needs for bonded warehousing, simple processing of goods and entrepot are also increasing. To meet the rising needs of foreign trade, China established its first Bonded Area known as Shanghai Waigaoqiao Bonded Area in June 1990 (China-shftz 2015). The Bonded Area refers to the area approved by the State Council, in which the Customs Administration adopts the special supervision measures, and goods can be stored within the Area for a relatively long time. The Customs Administration imposes closure management on the Bonded Area and bonded management on the overseas goods entering the Bonded Area. Goods entering the Bonded Area from other regions in China are deemed as exiting China (yet without tax refund). At the same time, the Bonded Area adopts the foreign exchange policies that are more preferential than those beyond the Bonded Area. Following the Bonded Area in Shanghai, China successively established 14 Bonded Areas in Tianjin, Shenzhen, Dalian, Qingdao, Guangzhou and Xiamen (Li Zhou 2014).

During this period, China was facing the major opportunity of world-wide industrial transfer, and its processing trade had the absolute advantages. As a result, the Export Processing Zone (EPZ) began to emerge. The EPZs mainly attracted processing enterprises whose raw materials came from abroad and whose products were exported. By providing ideal industrial infrastructures and exemption or reduction in tariff and income tax, the EPZs drew considerable investment. In 2000, the State Council approved the first 15 Pilot Export Processing Zones. Enterprises residing in these EPZs enjoyed preferential taxation and foreign exchange administration policies, thus greatly pushing forward the development of "processing on order." To date, China has established 57 EPZs. In the beginning, most EPZs mainly engaged in bonded processing. But today, they have developed into comprehensive regions that have such functions as bonded processing, logistics, testing, maintenance, and R&D (Customsnet 2011).

9.1.2.2 Period of Opening-Up Development

When the Bonded Areas were first set up in China, they were separated from the ports, resulting in the import or export goods having to go through clearance twice when entering or leaving the Bonded Area. To some large extent, this has impacted the operating efficiency in the Bonded Areas. To improve the operating efficiency in Bonded Areas and push forward regional economic growth, China has erected the direct channels between Bonded Areas and port areas so as to integrate their functions. In December 2003, Shanghai Waigaoqiao Bonded Logistics Zone was established as China's first Bonded Logistics Zone. After that, several Bonded Logistics Zones were also established in Dalian, Tianjin, Guangzhou and Qingdao.

To further advance the opening-up of its economy and integrate the functions of Customs special supervision areas, China began to set up Bonded Port Areas that possessed the functions of port of entry, logistics and processing in 2005. The policies in these areas integrated the policies of EPZ, Bonded Area and port. In December 2005, Yangshan Bonded Port Area was put into use as China's first Bonded Port Area. Up to now, China has 14 bonded port areas in Tianjin, Dalian and Qingdao.

To push forward opening-up of the inland economy, China has also set up the Comprehensive Bonded Zone in inland areas that has the functions of Bonded Port Area. The Comprehensive Bonded Zone carries out the taxation and foreign exchange policies of Bonded Port Area. With the functions of Bonded Area, Export Processing Zone, Bonded Logistics Zone and port, the Comprehensive Bonded Zone can engage in such business as international transfer, distribution, procurement, entrepot trade and export processing. In December 2006, the Suzhou Industrial Park was approved as the first Comprehensive Bonded Zone in China. According to the relevant data published on the websites of the General Administration of Customs and the China Free Trade Zone & Export Processing Zone Association, by February 2015, China had 41 Comprehensive Bonded Zones in cities including Xinzheng of Zhengzhou, Shenyang, Xi'an and Kashgar of Xinjiang.

9.1.2.3 Period of Mutual Promotion Between Internal and External Opening-Up

In order to explore the new trail for the deepening of reform and expansion of opening-up on a fuller scale, China (Shanghai) Pilot Free Trade Zone was set up in 2013. Different from the previous Customs Special Supervision Zone, institutional innovation is the core in the establishment of the Pilot FTZ. The main purpose of establishing the Pilot FTZ is to promote the transformation and upgrading of the economy, establish the regulatory system on cross-border investment and trade, complying with the requirements for internationalization and legalization, and foster the new advantages in international economic cooperation and competition.

By the end of November 2014, there had been more than 22,000 firms invested in Shanghai FTZ, including 14,000 newly-established enterprises and 2,114 foreign enterprises; this number is 10.4 times more than that in 2013. The overseas investment had been growing at a faster speed and 160 overseas investment projects had been finalized. The outbound investment made by China totaled nearly 3.8 billion USD. The corporate logistic cost was lowered by about 10 % and the time spent on Customs clearance for import was sped up by 41.3 %. China had established 6,925 free-trade accounts with 4.89 billion RMB in balance (Shanghai Municipal Development and Reform Commission 2014).

In March 2015, the Central Politburo of the Communist Party of China reviewed and approved the *Framework Plan for the China (Guangdong, Tianjin, Fujian) Pilot Free Trade Zone*, which marked the commencement of the second batch of Pilot FTZs. The *Framework Plan,* on the basis of identifying the new model of regional economic cooperation and highlighting features of local economic development, also deems institutional innovation as the core, just as Shanghai FTZ does. To be specific, it mainly advocates reforms and innovation in terms of transformation of government functions, innovation in management models, trade and investment facilitation, and building the business environment featuring legalization.

9.2 Major Results in Building Shanghai FTZ

Since Shanghai FTZ was established more than one year ago, it has made major progress of this phase in areas of foreign investment management system, trade supervision regulations, financial innovation systems, transformation of government functions, and legal protection. The results are expounded in the following five aspects.

9.2.1 Actively Reforming the Investment Management System

9.2.1.1 Exploring the Negative List Management Mode

The Negative List is the major innovation in foreign investment management system in Shanghai FTZ. By bearing upon the international norms, the pre-establishment national treatment is provisionally given to the foreign investment. Shanghai FTZ has issued the 2013 Negative List and its 2014 revised version. These two editions adopt the *Industry Classification of National Economy* as the classification standard. The number of special management measures in the Negative List, after the revision, has been changed to 139, reducing by 51 measures; or 17.4 % more open than previously. Classified by types, of the 139 special management measures, there are 110 restrictive measures and 29 prohibitive measures. Classified by industries, there are six measures for primary industries, 66 measures for secondary industries (including 46 measures for manufacturing industries) and 67 measures for tertiary industries. The 2014 Negative List gives foreign business more rights to participate in the construction of infrastructures and land development in the Pilot FTZs, thus offering new opportunities for overseas service enterprises to enter the Chinese market.

9.2.1.2 Adopting Measures to Expand the Opening-Up of Service Industry

Since its establishment, Shanghai FTZ has suspended or terminated certain restrictive measures against the investment access (apart from banking institutions, information and telecom service) in six major fields, i.e., finance, shipping, commerce, professional service, culture, and social service. In September 2013, the State Council announced the *Framework Plan for the China (Shanghai) Pilot Free Trade Zone*, in which 23 measures pertained to the widening of opening-up in the above six fields. In June 2014, Shanghai FTZ put out 31 measures for broadening the opening-up, including 14 new measures for the service industry, which mainly loosened investment restrictions in the commerce, logistics, accounting and medical fields.

9.2.1.3 Establishing the Investment Service Promotion System

Shanghai FTZ has established the new service promotion system, in areas of corporate investment management, registration, supervision and other government services. In terms of outbound investment, Shanghai FTZ adopts the Filing System for investing overseas. In terms of enterprise's setup and registration, it adopts the Registered Capital Subscription System, the Granting License before Issuing Permit Registration System and Enterprise Filing System while loosening the requirements for registration of registered capital. In terms of corporate supervision, it adopts the Annual Report Pronouncement System; in terms of government service, it has instituted the service model of multi-departmental approval.

At the same time, the departments directly subordinate to the Central Government are also actively exploring new service models. For instance, the Taxation Department has launched ten innovative measures to conduct pilot test in modernized tax collection and administration.

9.2.2 Overall Reforms in Trade Supervision

9.2.2.1 Improving Trade Facilitation Level with Innovative Supervision Models

The Administration of Quality Supervision, Inspection and Quarantine (AQSIQ), the Inspection and Quarantine Department, the General Administration of Customs, and the Ministry of Transport had all issued measures that supported Shanghai FTZ in trade facilitation.

In the first half of 2014, Shanghai Customs Administration launched, in batches, 14 supervision/service regulations in Shanghai FTZ, such as the "Entry First, Customs Declaration Later" System. It also simplified the documents accompanying Customs clearance. According to the Customs statistics, since the new clearance regulations were put into use, the average Customs clearance time in Shanghai FTZ has been shortened by 41.3 %, compared with that outside the Zone; the average export clearance time has been shortened by 36.8 % (Eastday 2015a).

Shanghai Inspection and Quarantine Department has synthesized several standards for inspection and quarantine industry in the FTZ, including innovative risk management of biological materials and products entering or exiting China, the entrepot goods' place of origin certificate management system, the Pre-inspection System for imported goods, and the inspection & quarantine line monitoring and management model, etc. These innovations in inspection and quarantine systems have shown evident results. For example, after the implementation of the Pre-inspection System, the duration of time from the goods entering China to passing the import clearance has been shortened by at least 50 % (Chinanews 2014).

Shanghai Maritime Authority has issued a number of new regulations about supervision and service, such as the innovative inspection model, the pollution control methods, and the supervision process of operations safety, which effectively enhance the Customs clearance efficiency. Besides, Shanghai Border Control and Inspection Department in cooperation with Shanghai Customs Administration and Shanghai Inspection and Quarantine Department has been simplifying the Customs inspection formalities so as to improve the processing capacity at the inspection windows.

9.2.2.2 Enhancing Coordinated Innovation of Port of Entry Departments and Improving Efficiency of Supervision

Cross-border supervision and management of passengers, goods and vehicles involve many departments and their efficiency depends mainly on how well these departments coordinate with one another. Shanghai FTZ has actively developed cross-departmental coordinated supervision to form the "Electronic One-Stop Clearance" mode. The new mode relies on the intelligent Customs information platform, on which enterprises can finalize the one-stop clearance via the online and self-service process while keeping abreast of the Customs clearance process at any time.

In December 2014, Shanghai FTZ put into trial use of the new mode, which is shown to give the enterprises full autonomy and reduce their clearance cost. For instance, Shanghai Chang Lian logistics Co., Ltd., as the opening user to debut the one-stop clearance service, applied to the Customs Office to import a batch of computer mother boards. After logging onto the client end of the new mode and finalizing the input process, the Customs Office information system automatically completed the review and approval process, automatically collected the taxes and fees and automatically released the goods. The entire clearance process was completed in only 4 min (Shanghaigov 2014).

9.2.2.3 Pushing Forward Construction of the International Trade Single Window

The "International Trade Single Window" has become the important informatization infrastructure and platform for all countries to participate in international trade and compete in the international market. In June 2014, Shanghai FTZ initiated the construction of International Trade Single Window. To date, Shanghai FTZ's Single Window has established and opened six functional modules and one FZT Special Service, to realize the online checking and release of ships out of port, and real-time checking of approval status, etc. In the future, Shanghai FTZ will push forward the online testing of functional modules such as the general-trade export goods, cross-border e-commerce and application for ships in/out of port. It will also extend the system from the waterway ports to airports, integrate data from banks, logistics and enterprises, and realize the Single Window service for transmission and distribution of trade data.

9.2.2.4 Promoting Trade Transformation and Upgrading and Supporting Development of New-Type Industries

Shanghai Entry-Exit Inspection and Quarantine Bureau (SHCIQ) has launched the service system for the development of emerging industries such as cross-border e-commerce. On the one hand, SHCIQ has set up the quality and safety tracing and supervision mechanism for the cross-border e-commerce products, and set up the pilot information support system which are linked with the service enterprises for cross-border e-commerce. On the other hand, it has also supported the development of industries in the Pilot FTZ such as biological medicine, maintenance of old mechanical and electrical products, and re-manufacturing. According to the industry features and business needs, it drafts relevant supervision measures to improve the management efficiency and encourage relevant enterprises to settle in the FTZ.

9.2.3 Exploring the Opening-Up and Innovation in the Financial Field

9.2.3.1 Building the Free-Trade Account System

The Free-trade Account System is an important exploration in promoting the facilitation of financing exchange, expanding the opening-up in the financial market, and safeguarding against risks. In the Free-trade Account, RMB and foreign currencies are managed according to the same rules. To promote the establishment of the Free-trade Account System, in May 2014, the People's Bank of China, Shanghai Head Office issued the *Rules for Implementation of Separate Accounting Businesses in China (Shanghai) Free Trade Pilot Zone (Provisional)* and the *Rules for Prudent Management of Risks Relating to Separate Accounting Businesses in China (Shanghai) Free Trade Pilot Zone (Provisional)*. These two *Rules* regulate the Separate Accounting Business and Prudent Risk Management. By setting up the separate accounting units, the financial institutions would build the risk separation shield from the domestic business, and then provide the Free-trade Account service. This marks the establishment of the Free-trade Account System which lends better risk management.

The Free-trade Account System provides the carrier for the Pilot FTZs to experiment with the financing exchange business. In February 2015, the People's Bank of China, Shanghai Head Office issued the *Circular on the Implementation Rules for Macro-prudential Management of Offshore Financing and Cross-border Capital Flows for Separate Accounting Businesses in the China (Shanghai) Pilot Free Trade Zone*, which expanded the scale and channel for business players to conduct overseas financing, and established the paths and management approaches to the capital account convertibility. At the same time, the Free-trade Account System also laid the foundation for the development of subsequent financial innovation business.

9.2.3.2 Promoting Cross-border Use of RMB

To help Shanghai become the global RMB Pricing and Clearing Center of RMB products, in February 2014, the People's Bank of China (PBC), Shanghai Head Office issued the *Notice of the Shanghai Head Office of the People's Bank of China to Promote Cross-border Use of RMB in the China (Shanghai) Pilot Free Trade Zone*, which continued to simplify the procedures of cross-border use of RMB in the current and direct investment accounts in the FTZ, clarified the size and scope of overseas borrowings in financial institutions and enterprises in the FTZ, and launched the two-way cross-border RMB funds pool, cross-border RMB receipt and payment for items under the current account, the RMB settlement of cross-border e-commerce and other innovative businesses. Besides, when expanding the cross-border RMB use, to foster risk prevention, Shanghai FTZ has launched the instrument called "macro-prudential policy parameters" for the first time, so that the PBC can adjust the size of overseas RMB borrowings on a dynamic basis to cater to the needs of credit control.

9.2.3.3 Pushing Forward Liberalization of Interest Rates

Market-oriented reform of interest rates is the important measure to improve the Socialist Market Economy System and ensure that market plays the deciding role in resource allocation. On March 1, 2014, Shanghai FTZ lifted the cap of interest rate for the small-sum foreign currency deposit for residents in the FTZ. After that, Shanghai FTZ became the first one in China that realized complete liberalization of interest rate of foreign currency deposit in China. It made the first step in liberalized pricing of liabilities products. Lifting control on the interest rate of foreign currency deposits is a key step towards interest rate liberalization, which will improve the financial market service system in Shanghai and upgrade the position of Shanghai as an international financial center.

9.2.3.4 Deepening the Foreign Currency Management Reform

Shanghai FTZ adheres to the streamlined administration and delegation of power to the lower levels in foreign exchange management, shifting the focus on administrative approval to the focus on monitoring and analysis, shifting the focus on micro-management to the focus on prudent management, so as to improve management efficiency.

In February 2014, the Shanghai Branch of the State Administration of Foreign Exchange instituted the *Implementing Rules for Foreign Exchange Control to Support the Construction of the China (Shanghai) Pilot Free Trade Zone*, which made explorations and innovations in two aspects, i.e., deepening the foreign exchange management reform and preventing the foreign exchange balance risks. First of all, in terms of deepening the foreign exchange management reform, it simplifies the review of receipt/sale or purchase/payment of foreign exchange documents, and the registration formalities of FDI foreign currency, improves the claims debt and clearing management of foreign currency, and improves the centralized operation and management of foreign exchange funds for headquarters of multinational companies and foreign currency funds pool. It proposes facilitation measures such as the Pilot Policy of Foreign Exchange Management for International Trade Settlement Centers and banks engaging in commodity derivatives counter trading to clients in the FTZ. Second, in terms of preventing the foreign currency balance risks, it has enhanced the statistical monitoring and early-warning analysis while strengthening the off-site monitoring and on-site inspection.

9.2.3.5 Developing the Finance Lease Business

Finance Lease, also called Equipment Leasing, is the comprehensive trading integrating financing and leasing-assets that involve the leaser, lessee and supplier. The leaser, according to the lessee's selection of leased items and the supplier, purchases the leased items from the supplier for use by the lessee, who will pay the rent. Compared with the traditional financing approaches such as bank credit, the Finance Lease has natural advantages in the corporate financial channels. Because of this, Finance Lease and relevant innovative financial products are becoming one of the channels of corporate financing.

The financial policies issued by Shanghai FTZ involve some unique policies about Finance Lease. These policies expand the scope of Finance Lease business and reduce the restrictions over Finance Lease so as to promote its facilitation. Besides, Shanghai FTZ also simplifies the many approval formalities of Finance Lease so as to foster its development.

At present, Shanghai FTZ has been one of the busiest regions of Finance Lease industry in China, its number of leasing companies and the total amount of registered capital rank No. 1 in China. By the end of November 2014, Shanghai FTZ had ushered in 264 Chinese and overseas Finance Lease companies and 314 project subsidiaries, with the total registered capital exceeding 80 billion RMB (Eastday 2015b).

9.2.4 Accelerating the Transformation of Government Functions

9.2.4.1 Establishing the Linked Management Framework Cored on Administrative Commission

Organizations in Shanghai FTZ that exercise the government functions include three levels. First, the Administrative Commission of China (Shanghai) Pilot Free Trade Zone; second, relevant departments in Shanghai dispatched by the Central Government, and offices dispatched in the FTZ by the Shanghai Industry and Commerce Administration, Shanghai Quality Supervision Administration, Shanghai Taxation Bureau and Shanghai Public Security Bureau; third, other relevant departments and the Pudong New Area Government in Shanghai. The administrative scope and synergy mechanism involved in the three levels are confirmed in the form of law set by the *Regulation on China (Shanghai) Pilot Free Trade Zone.* Furthermore, supplementary measures are also enacted to promote sharing of supervision resources among departments so as to improve the capability of joint supervision and synergistic service.

9.2.4.2 Exerting Multi-partite Forces and Strengthening the Supervising Mechanism of the Market Players

Shanghai FTZ has instituted a series of measures to mobilize the multiple forces to participate in the supervision of price monopoly, market centralization and national security concerns over FDI. In October 2014, the *Measures for Law Enforcement against Monopolistic Agreements, Abuse of Dominant Market Positions, and Administrative Monopoly in China (Shanghai) Pilot Free Trade Zone* was issued. Besides, Shanghai FTZ has launched the supplementary measures which focus on encouraging the parties to participate in the anti-monopoly investigation so as to prevent centralized operators, monopoly agreement and abuse of market dominant position, etc.

9.2.4.3 Building Up Credit System Based on Information Swapping and Publishing

Shanghai FTZ has issued a number of regulations on corporate information management and publishing, so as to regulate the establishment of credit system. At present, Shanghai FTZ has initially established two systems, namely, the primary credit data and the credit award & penalty mechanism. The Primary Credit Data, by swapping with the Shanghai Public Credit Information Service Platform, the credit data of relevant judiciary departments and public facilities departments, realizes the integration of primary credit information of enterprises and individuals. The Credit Award &Penalty Mechanism fulfills the role of credit means and makes concurrent and ex post supervision over the commercial and social activities of enterprises or individuals, thus guiding and encouraging the individuals and enterprises to be honest and self-disciplined.

By the end of June 2014, Shanghai FTZ had performed credit inspection on 3,924 key enterprises, 1,172 Customs-supervised enterprises and 15,000 corporate senior executives in the FTZ; 9,747 enterprises had submitted annual reports (MOFCOM 2014). In February 2015, Shanghai FTZ, on the basis of credit information swapping, opened the Credit Information Comprehensive Inquiry Platform, thus realizing service functions such as public credit information inquiry and financial credit information inquiry and sharing.

9.2.5 *Innovating the Systems of Policies, Laws and Regulations*

9.2.5.1 Ministries and Commissions in China Issuing Various Documents to Support the Establishment of Shanghai FTZ

The Ministry of Industry and Information Technology, the Ministry of Finance, the Ministry of Transport, the Ministry of Culture, the Ministry of Justice, the People's Bank of China, the General Administration of Customs, the State Administration of Taxation, the Administration for Industry, the Administration of Quality Supervision, Inspection & Quarantine, the China Banking Regulatory Commission, the China Securities Regulatory Commission, and the China Insurance Regulatory Commission have issued a series of supportive documents in financial and investment management, trade supervision, and construction of shipping centers, such as the *Notice of the China Banking Regulatory Commission on Issues concerning Banking Supervision in China (Shanghai) Pilot Free Trade Zone,* and the *Implementation Opinions of the Ministry of Transport and the Shanghai Municipal People's Government on Actualizing the Framework Plan for China (Shanghai) Pilot Free Trade Zone and Accelerating the Construction of Shanghai International Shipping Center.*

9.2.5.2 Establishing Special Management Measures for Adjustment of Related Regulations, Documents and Departmental Rules

In order to perfect the legal assurance system and adapt to further broadening of the opening-up, the State Council resolved that Shanghai FTZ has the right to temporarily adjusted related administrative laws and regulations in accordance with the relevant provisions. For example, the FTZ can adjust some laws and regulations contained in the *Law of the People's Republic of China on Foreign-Capital Enterprises* and the *Law of the People's Republic of China on the Protection of Taiwan Compatriots' Investment*. Additionally, the State Council stressed that Shanghai FTZ shall strengthen its local legislation; for pilot reforms which had reached the maturity stage, the related regulatory documents can be updated as local laws and regulations.

9.2.5.3 Issuing Local Comprehensive Laws and Regulations of the FTZ

Shanghai has issued a series of local comprehensive regulations like the *Special Management Measures (Negative List) for Foreign Investment in China (Shanghai) Pilot Free Trade Zone* to boost the construction of free trade zone.

On August 1, 2014, China's first local regulation about FTZ, the *Regulation on China (Shanghai) Pilot Free Trade Zone*, initiated its trial implementation. This

Regulation specifies across-the-board matters concerning investment, trade, financial, tax, supervision and the construction of legal framework of the pilot free trade zone.

9.3 Outline of the Construction Plan of the Guangdong, Tianjin and Fujian FTZs

Guangdong, Tianjin and Fujian FTZs are China's second batch of free trade zones. These three FTZs replicated many successful experience of Shanghai FTZ, in several aspects such as accelerating the changes of government functions, further opening the investment field, and promoting innovation of the financial systems and trade regulatory system. Meanwhile, they also each has distinct geographical characteristics and positioning. The focus and characteristics of the construction plan of Guangdong, Tianjin and Fujian FTZs are each described as follows.

9.3.1 The Guangdong Free Trade Zone

According to *the Overall Plan for China (Guangdong) Free Trade Zone* and the relevant contents of the official website of Guangdong FTZ, the total lands area of Guangdong FTZ is 116.2 km^2, including the Nansha Area of Guangzhou, Qianhai and Shekou Area of Shenzhen and Hengqin Area of Zhuhai. Its main construction goal is to build up Guangdong FTZ to become a cooperation demonstration zone of Guangdong, Hongkong and Macao, an important hub of the twenty-first century Maritime Silk Road (One Road), and the pioneering ground of the country's new round of reform and opening-up.

9.3.1.1 Exploring New Mode of Economic Cooperation Between Guangdong, Hong Kong and Macao

To explore the new mode of economic cooperation between Guangdong, Hong Kong and Macao, Guangdong FTZ will make certain innovations mainly in the following three aspects. (1) In promoting the liberalization of the service trade in Guangdong, Hong Kong and Macao, it will abolish or relax the investment restrictions in fields of finance, trade and other services, innovate the system mechanism facilitating the convenient flow of service elements. (2) In enhancing international shipping services, it will create a regional logistics hub, promote the shipping finance and business innovation, and promote the development of shipping industry. (3) In the financial field, it will expand the cross-border RMB business, facilitate investment and financing, and reinforce the risk prevention system.

9.3.1.2 Fortifying Its Radiating and Commanding Functions

Commanding the industrial upgrading of the Pan-Pearl River Delta (PPRD) and the inland region will be the key content of the Guangdong FTZ construction program. The main measures include the following. (1) Effect the high-end elements gathering advantages of the FTZ; build the technology research and development, the intellectual property and other public service platforms; bring about all types of service businesses. (2) Support the extensive economic and trade cooperation between the FTZ and PPRD, create a comprehensive service area in PPRD, and promote regional industrial development. (3) Rely on the advantages of Hong Kong and Macao in financial services, information and news, international trade network, and risk management, construct the international communication platform for mainland enterprises and individuals, to further enhance the radiation and driving functions of the FTZ.

9.3.1.3 Optimizing the Business Environment

The Guangdong FTZ program proposed the key measures of constructing an international, market-oriented and legalized business environment, from the three angles of optimizing the legal environment, innovating the administrative management system, and building a market access and supervision system of "lenient-entry and tight-monitoring." Compared with Tianjin and Fujian FTZs, Guangdong FTZ highlights the system innovation in enhancing the capability of administrative management. For example, it innovatively proposed to establish an administrative advisory system, establishing a professional advisory committee composed of professionals from Guangdong, Hong Kong and Macao, to give advice on the development of the FTZ.

9.3.2 The Tianjin Free Trade Zone

According to *the Overall Plan for China (Tianjin) Free Trade Zone* and the relevant contents of the official website of Tianjin FTZ, the total land area of Tianjin FTZ is 119.9 km^2, including the Tianjin Port Area, Tianjin Airport Area and Tianjin Binhai CBD Area; the three large areas all have their own characteristics and complementary advantages. Tianjin FTZ serves two major strategies: the Beijing-Tianjin-Hebei Coordinated Development and the "One Belt and One Road." It will highlight its own characteristics in the construction process, and achieve greater breakthroughs in fields such as the development of substantive economy and the growth of financial leasing industry.

9.3.2.1 Improving the Level of Financial Leasing Development

The scale of financing lease industry of Tianjin port accounts for nearly 1/4 of the nation's financial leasing industry. Tianjin FTZ will make policy system innovation of leasing industry on the existing basis, and form a leasing industry development environment of international level. Substantively, in terms of constructing the leasing service system, the FTZ will construct a rental service platform, unify the internal and external approval processes and access standards for financing lease enterprises and other institutional innovation. In terms of supervision, it will implement the mutual recognition of Customs control policy according to the actual logistics needs of the financial leasing enterprises.

9.3.2.2 Seaport-Airport Intermodality Development

Having seaport and airport co-located in the FTZ is a major feature of Tianjin FTZ. Improving the collection/dispersion/distribution system, exploring the seaport-airport intermodality operations mode and developing supplementary services will be the system innovation focus of the Airport Area of Tianjin FTZ. At present, the Airport Area has set up the Tianjin Airport Economic Zone Investment Assurance Center, and four industry funds and four special project funds (i.e. high-end manufacturing, modern service industry, high-tech industry and talents) to promote the innovation and development of the airport area.

9.3.2.3 Serving the Beijing-Tianjin-Hebei Coordinated Development

Tianjin FTZ aims at fulfilling the following five major reform and innovation tasks: accelerating the Beijing-Tianjin-Hebei Coordinated Development, of the region, expanding the opening-up of investment fields, promoting the transformation and upgrading of trade, deepening the opening-up and innovation in the financial field, and accelerating the transformation of government functions. In regard to coordinating the development of the Beijing-Tianjin-Hebei region, Tianjin FTZ will make substantial construction in four aspects, i.e., enhancing the port's service radiation function, promoting regional industrial transformation and upgrading, pushing forward regional financial market integration, and elevating the scientific and technological innovation, so as to build a high level opening platform.

9.3.2.4 Promoting the Construction of the "One Belt and One Road"

Tianjin is an important node in the "One Belt and One Road" strategy, with industrial agglomeration advantages of the Beijing-Tianjin-Hebei region and the location advantage of neighboring Japan and South Korea. In order to promote the construction of the "Belt and Road," Tianjin FTZ will (1) build the cross-border logistics center, carry out high-end logistics services, to achieve differentiated, intensive development; (2) cultivate more comprehensive logistics service providers, to expand the sea, land, air freight forwarding services and multimodal transport forwarding services; (3) optimize the layout of inland port, duplicate and extend the system innovation achievements of FTZ, to enhance the entrepot trade service functions towards countries and areas along the Belt and Road.

9.3.3 The Fujian Free Trade Zone

According to *the Overall Plan for China (Fujian) Free Trade Zone* and the relevant content of the official website of Fujian FTZ, the total land area of Fujian FTZ is 118.04 km^2, including the Pingtan Area, Xiamen Area and Fuzhou Area, each area with its own characteristics. Fujian FTZ's main objectives are to explores new models of economic cooperation between Fujian and Taiwan, and to strengthens the exchange and cooperation with countries and regions along the twenty-first century Maritime Silk Road (One Road), so as to accumulate experience for China's opening-up.

9.3.3.1 Promoting Investment and Trade Freedom with Taiwan

To create the free flow of service factors between Fujian and Taiwan, and a deeper opening-up to Taiwan in service trade, Fujian FTZ will further expand its opening to Taiwan in the field of telecommunication, transportation, tourism, health care, professional technology and other services. It will also actively explore new modes of industrial cooperation between Fujian and Taiwan, to effect the deeper level integration of industrial chain between Fujian and Taiwan. In order to promote the free flow of cross-strait personnel and goods, it will explore and implement more convenient immigration policy and innovative regulatory model for Taiwan residents.

9.3.3.2 Giving Priority to Pilot-Testing Cross-Strait Financial Cooperation

To deepen the economic cooperation across the Strait, cross-strait financial cooperation became the major content of Fujian FTZ, which mainly includes the measures on in the following three aspects: (1) further opening the financial service industry to Taiwan-funded businesses, e.g., exploring the measures for lowering the access and business threshold for Taiwan-funded institutions; (2) expanding the scope of financial business, e.g., exploring the measures for RMB loaning business between banking institutions in pilot free trade zones and their Taiwan counterparts, and permitting operating units of Taiwan banks the freedom to conduct RMB business once setting up in the FTZ; (3) relaxing the restrictions on investment, loosening or removing the restrictions on equity ratio of Taiwan-funded financial institutions participating in the securities and funds institutions in the FTZ.

9.3.3.3 Fostering New Advantages of Pingtan in Opening-Up and Development

The main content of developing Pingtan Area is to build up a common homestead and an international tourism island. In order to develop new advantages, Pingtan Area will give priority to seeking institutional system innovation on in the following three aspects: (1) facilitating the interchange of fund and personnel for trade investment; (2) pushing forward shipping liberalization, try to the simplifying the port entry and exit procedures of ships and implementing mutual recognition of supervision with Taiwan; (3) building up an international tourism island, implementing internationally accepted tourism service standards, accelerating the transformation and upgrading of tourism industry, and instituting a more convenient entry and exit system for Taiwan residents.

9.3.3.4 Pushing Forward the Construction of the Kernel of the "One Road"

Fuzhou City is a critical node on the Maritime Silk Road ("One Road") in the twenty-first century. Fujian FTZ will expand the exchange and cooperation with the countries and regions along the One Road, and build up the kernel of One Road and a new platform for cooperation and exchange. The pilot FTZ will emphasize the innovation in Customs clearance mechanism, support the cooperation and exchange among the countries and regions along the One Road in regard to Customs, inspection and quarantine, certification and recognition, standards and metrology. It will also explore new approaches for safe and convenient cooperation in trade supply chain among the countries and regions along the One Road.

9.4 Development Status and Prospect of Logistics in the Free Trade Zones

Construction of pilot free trade zones necessitates the support of logistics system, and conversely, pilot free trade zones may also play a radiating and driving role for regional economic development by virtue of the supply chain and logistics system. At present, the logistics industry is developing swiftly and its service functions being improved continually in Shanghai Pilot Free Trade Zone. Seen from the construction plans promulgated by Guangdong, Tianjin and Fujian Free Trade Zones, the future construction and development of pilot free trade zones will effectively boost the transform and development of the logistics industry, drive the innovation of logistics service, improve the international competitiveness of logistics enterprises, and accelerate the progress of integrated regional logistics.

9.4.1 Current Development Status of Logistics in Shanghai Pilot Free Trade Zone

9.4.1.1 Current Status of Throughput of the Airports and Sea ports

At present, the international competitiveness of Port of Shanghai is continually being improved; airport throughput is growing faster than that of sea port. With respect of to sea port, the container throughput of Port of Shanghai was 35.29 million TEUs in 2014, up by 5 % year-on-year, and ranking No. 1 in the world (Askic 2015). The development of port business imposed higher requirements on the capability of Port of Shanghai. The construction of automated terminal of Phase IV of Yangshan Port was inaugurated on December 23, marking a significant step toward the construction of an automated terminal. With respect to airport, Pudong International Airport achieved a passenger throughput of 51.69 million person-times in 2014, increased by 9.53 % year-on-year, and ranking 20th in the world; cargo and mail throughput topped 3.18 million tons, grew by 9.28 % compared with that of the previous year, and ranking third in the world (Eastday 2015c). In 2015, the fourth runway of Pudong International Airport was put into operation, making Pudong the first domestic airport which has four runways, providing a larger operating space for the parallel approach mode implemented by the Air Traffic Management Bureau.

9.4.1.2 Construction of Shipping Center Achieving Preliminary Results

(1) Innovation of Ship Registry and Ship Management Policy

Shanghai FTZ has reformed the international ship registry system, relaxed the conditions on ship registry, age limits, crew staffing and ship registry types. With respect to ship management, it has relaxed the market access and share ratio limits for international ship management firms. These measures have boosted the development of intentional ship management firms. As of the end of August 2014, three foreign-funded ship management enterprises had settled there (Yicai 2014).

(2) Rapid Development of Transshipment in the Pilot Free Trade Zone

Based on the implementation of export tax rebate program in Shanghai FTZ and the loosening of cabotage of domestic shipping market along the coast, Shanghai Customs promulgated the operations scheme for transshipment and consolidation in November 2014. Export and import sorting and consolidation business of freight along the river, coast, near-sea and ocean was carried out within the Customs area under special supervision. This further accelerated the development of container transshipment. In 2014, the proportion of container waterway-waterway transit was 45.8 % in Port of Shanghai, an increase of 0.4 percentage points year-on-year; the proportion of international transit was 7.1 %, up by 0.1 percentage points (21PI 2015).

(3) Rapid Development of High-end Shipping Service

Shipping finance and insurance services attained rapid development in Shanghai FTZ, and the rules for maritime trial and arbitration were also further perfected. Shanghai established the first shipping and finance industry base, which attracted high-end institutions on in the shipping and finance industry chain to settle down there. As of December 2014, a number of financial leasing companies had established 151 subsidiaries in the FTZ (Southmoney 2015). In addition, the FTZ relaxed the pilot development of the articles of Shanghai Institute of Marine Insurance, abolished the prior-approval for setting up of subsidiaries of Shanghai-based shipping insurance operations centers and reinsurance companies in the FTZ, and canceled the prior-approval of employment qualification of senior executives of insurance subsidiaries in the FTZ. These measures jointly energized the development of shipping insurance business in the FTZ. By the end of 2014, approximately 50 % hull insurance companies around China had settled in Shanghai, and 80 %–85 % ocean insurance companies gathered in Shanghai (Eastmoney 2015).

In regard to maritime justice and arbitration rules, Shanghai FTZ implemented the Arbitration Rules of the China (Shanghai) Pilot Free Trade Zone, and Shanghai Maritime Court established the Free Trade Zone Court in April 2015; thus maritime justice and arbitration rules were further improved.

9.4.1.3 Agglomeration of Supply Chain and Logistics Operations Centers

As early as 2008, 491 of the World Top 500 Enterprises had settled in Shanghai. In order to facilitate Shanghai FTZ to establish an operations center integrating trade, logistics, settlement, payment and other functions, the FTZ hastened the establishment of an international trade settlement center, to assist enterprises streamline the settlement process and provide a sound environment for global supply chain management for enterprises. It is conjectured that most Asian-Pacific supply chain coordinating centers of the World Top 500 Enterprises will move from Hong Kong and Singapore to Shanghai, and this will accelerate the aggregation of supply chain logistics and logistics operations centers of transnational companies in Shanghai (Chinanews 2013).

9.4.1.4 Rapid Development of Logistics Value-added Services of Bulk Commodities

In order to effectively improve the transaction quotation capability of Chinese institutional investors in the international bulk commodity market, Shanghai FTZ published the *Interim Provisions on the Administration of Trading in Bulk Commodities on Spot Markets in China (Shanghai) Pilot Free Trade Zone* in November 2014, and successively approved the establishment of eight commodity transaction centers in 2015.

On this basis, trading of bulk commodities became ever-more active in the FTZ, and new services integrating trade, logistics, settlement, futures, financing and investment emerged constantly. These services support the expansion of the pilot varieties, the enterprises and the scope of bonded futures delivery, and expand the continuous development of warehouse receipt pledging and other high-end logistics services. For instance, Shanghai Clearing House launched the business in Central Counterparty (CCP) clearing, the CNY iron ore swap, and the CNY steaming coal swap.

In addition, Shanghai FTZ also achieved rapid development in high-end services for imported car logistics. The FTZ actively carried out the program for parallel imported cars, innovated auto logistics services, extended the imported car supply chain; resulted in a 9.1 % increase in the auto ro-ro business, keeping Shanghai's leading position in China (Unbank 2015).

9.4.1.5 Rising of Cross-border E-commerce Logistics

In 2013, Shanghai FTZ established the first cross-border e-commerce platform in China, and entered into a strategic cooperation agreement with Amazon in 2014, bringing in global products by bonded import or direct mail delivery from abroad to engage in import business. As of April 2015, the total value of Amazon's direct-buy import orders had hit 31.85 million RMB, accounting for more than 90 % of the total value of cross-border e-commerce orders (Yicai 2015). Moreover, the business of "New-Tao" borne by the international courier leader DHL for delivery was kicked off in Shanghai FTZ.

The institutional innovation of Shanghai FTZ in cross-border e-commerce goods clearance, settlement and payment will boost the development of cross-border e-commerce; overseas warehouse, overseas distribution center, and relevant businesses will also attain vigorous development.

9.4.2 Outlook on the Logistics in the Pilot Free Trade Zones of Guangdong, Tianjin and Fujian

9.4.2.1 Outlook on the Logistics in the Pilot Free Trade Zone of Guangdong

In recent years, with the industry optimization and upgrade and the development of export-oriented economy, the regional logistics industry in the Pearl River Delta Region exhibits a trend of transformation toward high-end logistics. A high-end logistics industry agglomeration zone was established in Qianhai of Shenzhen, which attracted a large number of foreign investments in the logistics industry and participation in the construction of logistics infrastructures. In addition, Guangdong Province has also continuously strengthened the docking with Hong Kong and Macao in urban planning and transportation network, e.g., construction of HK-Zhuhai-Macao Bridge laid a solid foundation for the cooperation and development of the logistics industry of Guangdong, Hong Kong and Macao.

With the construction of Guangdong Free Trade Zone, the logistics industry will embrace a new development opportunity. First, the transportation and shipping service market will be opened wider to the service providers of Hong Kong and Macao, who will bring about international management standards and rules for the Mainland's international shipping market; this will provide room for the development of logistics enterprises in the Free Trade Zone. Secondly, Guangzhou Free Trade Zone will further facilitate the coordinated development of the sea ports and air ports in Guangdong, Hong Kong and Macao, and jointly build up the logistics hub for the "One Road," and boost the extension of international high-end shipping industry of Hong Kong and Macao to the Mainland. Finally, Guangdong, Hong Kong and Macao will further the prosperity of cross-border e-commerce, settlement, shipping finance and insurance, and other high-end logistics businesses of the three regions.

Guangzhou Free Trade Zone will also enhance the agglomerating and radiating functions of Pan-Pearl River Delta by building up a comprehensive service area for the development of Pan-Pearl River Delta region. It will carry out bulk commodities spot transaction and international trade, explore the establishment of international commodity transaction center, information center and pricing center, and will, by interaction of domestic and overseas trades, facilitate the integration of logistics industry and trade finance. In addition to advancing the high-end logistics industry, it will further elevate the radiating service capacity of logistics industry.

9.4.2.2 Prospect for Logistics Development in Tianjin Free Trade Zone

Tianjin is an important port city in Northern China, in which the shipping and logistics industry plays an important role. Through many years of development, Tianjin fully leverages its regional advantage, and continues to extend its radiation scope of port and waterway logistics toward the northwest hinterland along the "One Belt;"

the extent of fusion of its logistics industry and high-end manufacturing industry is deepening continually, and the international logistics network system is improving day by day.

By fully utilizing its regional advantages and centering on the further reform and opening-up of the Coordinated Development of the Beijing-Tianjin-Hebei region, the development of Tianjin FTZ's logistics industry will have the following characteristics. (1) The Beijing-Tianjin-Hebei regional Customs clearance integration reform, the construction of bonded logistics network and the construction of the logistics park exclusive to Beijing and Hebei areas will further promote the regional logistics integration of Beijing-Tianjin-Hebei area. (2) The development of professional port and waterway logistics and bonded logistics will further promote the innovation of Beijing-Tianjin-Hebei integration logistics services. (3) The Coordinated Development of the Beijing-Tianjin-Hebei region will bring about the prosperity of the regional air logistics market; the freight service network of Binhai International Airport will be further improved, thus driving the sea-air intermodality development.

Tianjin FTZ is an important node of the "One Belt and One Road," and is located in the North inland port network system with superior location advantage. It can be predicted that the intermodal transport system will attain rapid development in the future, the internationalized service level of logistics will be further improved, and the logistics service centering on cross-border e-commerce business will gain further momentum.

9.4.2.3 Prospect for Logistics Development in Fujian Free Trade Zone

Fujian province is situated at the junction of the southeast coastal region and the eastern coastal region in China, and at the center of Western Economic Zone across the Taiwan Straits. The logistics industry shows the development pattern of modern logistics industry network, basing on ports and focusing on central cities of Fuzhou and Xiamen.

Fujian FTZ will promote the cross-straits economic and financial cooperation, and make the cross-straits trade more convenient and effective. It can be predicated that Fujian FTZ will become the logistics and supply chain operations center across the straits, with its radiation capacity strengthened and regional logistics further developed. By virtue of the reform of financial system, the cross-straits cooperation will be reinforced and the cross-straits logistics service mode will continue to innovate to form a cross-straits logistics hub with a broader scope.

The construction of Fujian FTZ will also integrate the regional logistics resources, and drive the development of regional logistics. The development of Customs clearance facilitation and trade liberalization will promote the continual improvement of bonded area supervision service, bonded area logistics network and the comprehensive bonded area network; it will also enhance the regional radiation capacity of Fujian FTZ, making the pilot free trade zone an important logistics node of the Maritime Silk Road.

9.4.3 The Effect of Free Trade Zones on China's Logistic Industry

9.4.3.1 Promoting the Transformation Development of the Logistics Industry

Changes in development environment dictate the logistics enterprises to continually improve the degree of specialization, expand the spatial service scopes, integrate the logistic functional elements, logistics, and innovate the service mode. FTZ's opening-up to the service industry will bring about the rapid development of international ocean shipping operations, international marine cargo handling, international marine container station and storage yard businesses, as well as the related logistics value-added services. This will drive the development of specialized market segments, and thus reduce the operating costs of enterprises engaged in distinctive logistics. Meanwhile, with the reform in investment system, the investment environment of logistics enterprises will be improved; this will further promote the merger and reorganization of enterprises, realizing regionalized and networked operations on a larger scale. The "Measures for the Examination of Concentrations of Business Operators" implemented in FTZs will afford more perfected market competition environment for the logistics industry, promote the logistic enterprises to more efficiently integrate logistics services, expand the breadth and depth of supply chain services, thus expediting the transformation development of logistics services.

9.4.3.2 Driving Innovation in Logistics Service

With the construction of FTZ infrastructures and the improvement of market environment, FTZs in China will relentlessly explore the innovation in logistics service. International commodity exchange and resources allocation platform will be built in FTZs to carry out international trade in energy products, basic industrial raw materials and bulk agricultural products, expand the pilot sites of futures delivery in bonded area, expand warehouse receipt financing and other functions. It will also promote the conglomeration of cross-border trade in information flow, logistics and capital flows, accelerate the in-depth integration of trade and logistics, thus pushing forward the continual innovation in services in the logistics industry. In the future, with the development of international division of labor and trade globalization, emerging enterprises in supply chain finance, supply chain service platform and cross-border e-commerce logistics will gradually amalgamate. Meanwhile, with the development of internet technology, internet will further fuse with finance, trade and logistics, which will further the prosperous development of the logistics industry.

9.4.3.3 Efficiently Elevating the International Competitiveness of China's Logistics Enterprises

With the reform in FTZ's trade supervision, the construction of free-trade account, the cross-border use of RMB, the reforms in foreign exchange management system, as well as the continuing innovation in talent development, the comprehensive competitiveness of Chinese logistics enterprises will be improved. In the reform of trade supervision, the Customs, the Inspection and Quarantine Bureau, the maritime authority and other departments have adopted many systems and measures to innovate the supervision service mode, and accelerated the interconnection and interworking among various supervision institutions. These actions have promoted the free flow of goods, services and other elements in FTZs, reduced the average Customs clearance time and costs for import and export trade. The innovation in financial system will reduce the financing cost of logistics enterprises and provide convenience for outward-bound investments by logistics enterprises, thus enhancing the international competitiveness of Chinese logistics enterprises. Meanwhile, the FTZs will also encourage the collaborative innovation of scientific research institutions, colleges and universities, and enterprises in the heartland of China, and accelerate the cultivation of international logistics talents. This will provide the assurance of high-end talents for logistics enterprises in elevating their international competitiveness.

9.4.3.4 Accelerating the Progress of Regional Logistics Integration

The construction and development in the Beijing-Tianjin-Hebei region, the Yangtze River Economic Belt, the Pearl River Delta and the "One Belt and One Road" will help to promote the regional economic integration in the areas where these policies are implemented. With the continual innovation in trade, finance and other supervision systems in these areas and the continuing extension of the supply chain operating space of logistics enterprises, a unified logistics market will take shape in these areas. Meanwhile, the nodes and key hubs of high-end regional logistics will gradually emerge. The construction of grand regional logistics channels and node network will also provide an ever-expanding market for the logistics enterprises that centered on international logistics and engaging in the whole-process logistics supply chain services.

9.5 Summary

Free Trade Zone is the measure adopted by China to push forward reform and opening-up under the unoptimistic overseas economic and trade conditions, and the circumstance that domestic economy urgently needs transformation. FTZ has profound meaning in spurring the vitality of China's economy and promoting its economic transition. This Chapter first presents the development trail of China's special economic zones and the background for the birth of China's pilot free trade areas. It then succinctly introduces the important phased objectives and replicable experience achieved by Shanghai FTZ in investment management, trade supervision, financial revolution, transformation of government functions, and legal system construction. On this basis, the highlights of the respective plans for the construction of Guangdong, Tianjin and Fujian free trade zones are outlined. Guangdong FTZ emphasizes on promoting the in-depth economic cooperation between Mainland China and Hong Kong & Macau; Tianjin FTZ is based on promoting the Beijing-Tianjin-Hebei Coordinated Development; Fujian FTZ focuses on deepening the cross-straits economic, financial and service cooperation. Finally, the Chapter briefly analyzes the status and results of Shanghai FTZ's logistics development, and preliminarily expounds on the effect of FTZs on future development of China's logistics industry by summarizing the construction plans of Guangdong, Tianjin and Fujian FTZs.

References

21PI (2015) In 2014, the cargo throughput of port of Shanghai reached 755.29 million tons. 21PI. http://www.21pi.cn/news/show.php?itemid=5743. Accessed 3 Mar 2015

Askic (2015) The ranking of 2014 world's 10 largest container port throughput. Askic, http://www.askci.com/chanye/2015/02/05/104043yjlu.shtml. Accessed 5 Feb 2015

China's Foreign Trade Situation (2014) Ministry of Commerce of the People's Republic of China

Chinanews (2013) Shanghai will become the Asia-Pacific supply chain hub. Chinanews. http://finance.chinanews.com/cj/2013/09-16/5287068.shtml. Accessed 16 Sep 2013

Chinanews (2014) Shanghai-FTZ carry out pre-inspection system. Chinanews. http://finance.chinanews.com/cj/2014/05-09/6155293.shtml. Accessed 9 May 2014

China-shftz (2015) Brief introduction of Waigaoqiao bonded area, China-shftz. http://www.china-shftz.gov.cn/wgqIndex.aspx

Customsnet (2011) Promote the development of special economic zones. Customsnet. http://qingdao1.customs.gov.cn/publish/portal105/tab45056/. Accessed 29 Nov 2011

Eastday (2015a) Customs clearance time was reduced by 41.3 %. Eastday. http://finance.eastday.com/m/20150131/u1a8562809.html. Accessed 31 Jan 2015

Eastday (2015b) FTA finance lease should give more support to small enterprises. Eastday. http://finance.eastday.com/m/20150323/u1a8634211.html. Accessed 23 Mar 2015

Eastday (2015c) China Civil Aviation report. Eastday. http://gov.eastday.com/qxxc/node39/node43/node76/u1ai21761.html. Accessed 13 Apr 2015

Eastmoney (2015) Shipping insurance companies are being concentrated in Shanghai. http://finance.eastmoney.com/news/1355,20150130473394546.html. Accessed 30 Jan 2015

Li Zhou (2014) Study of China bonded area. Commercial. May 2014

Li Chunding, Zhao Meiying (2015) Changes in international trade rules and China's countermeasures. Chinese market. May 2012

MOFCOM (2014) Results and replicated path of Shanghai FTZ. Chinese Academy of International Trade and Economic Cooperation. http://www.caitec.org.cn/article/xsyj/wz/201412/321.html. Accessed 30 Oct 2014

Shanghai Municipal Development &Reform Commission (2014) Achievements of reform. Shdrc. http://www.shdrc.gov.cn/main?main_colid=361&top_id=316&main_artid=25479. Accessed 30 Nov 2014

Shanghaigov (2014) Independent declaration of duty and taxes is being piloted. Shanghaigov. http://www.shanghai.gov.cn/shanghai/node2314/node2315/node15343/u21ai962179.html. Accessed 16 Dec 2014
Southmoney (2015) Finance lease of Shanghai FTZ ranked first in China. Southmoney. http://www.southmoney.com/hkstock/ggxinwen/201501/252062.html. Accessed 5 Jan 2015
Unbank (2015) Unbank, Shanghai International Port (Group) Co., Ltd.. http://www.unbank.info/page?sid=1&pid=229&fid=543&item=1160073. Accessed 28 Mar 2015
World Economic Outlook (October 2014) International Monetary Fund
Xinhuanet (2014) WTO promote trade facilitation agreement. Xinhuanet. http://news.xinhuanet.com/world/2014-11/28/c_1113444045.htm. Accessed 28 Nov 2014
Yicai (2014) The State Council: Shanghai FTZ steadily carry out foreign-owned ship management firm pilot. Yicai. http://www.yicai.com/news/2014/09/4015162.html. Accessed 3 Sep 2014
Yicai (2015) Shanghai FTZ: Amazon direct purchased import orders breaking 100,000. Yicai. http://www.yicai.com/news/jrtt/2015/04/4608258.html. Accessed 20 Apr 2015

Chapter 10
Transformation and Upgrading Model of China's Port and Its Outlook

Yong Liu

Port is well-recognized as the crucial infrastructure for a country and the major hub in its comprehensive transport system. China has become globally notable in port. Due to the enormous change in the environment of the Chinese and overseas economic and social development, China's Central Government and local governments have issued a number of policies to promote the transformation and upgrading of the nation's ports. Ports in China have striven to explore suitable models for port transformation and upgrading. This chapter includes three sections. Section 10.1 introduces the macro-economic background, industrial background and policy background of China's port transformation and upgrading. Section 10.2 describes five transformation and upgrading models of China's ports, and Section 10.3 presents an outlook on the future trends of China's port transformation and upgrading, taking into consideration of China's future development strategy and technological advancement.

10.1 Background of China's Port Transformation and Upgrading

Conditions and environment for the development of China's ports are changing fundamentally. In recent years, throughput in China's ports has down-shifted from the rapid growth to the new norm featuring steady growth. Large ships are raising new requirements for the port's equipment and its operations and service capability. Development of China's ports is in urgent need to expand the service functions and to change the traditional mode of growth. To promote port transformation and upgrading, the Central Government and local governments in coastal areas have issued various related policies.

Y. Liu (✉)
Business School, Nankai University, Tianjin, People's Republic of China
e-mail: liuyongth1@sina.com.cn

L. Wang et al. (eds.), *Contemporary Logistics in China*, Current Chinese Economic Report Series, DOI 10.1007/978-981-10-1052-1_10

10.1.1 Macro-economic Background of China's Port Transformation and Upgrading

10.1.1.1 Port Development Model Featuring Size Expansion No Longer Sustainable

Due to the decade-long vigorous development, throughput capacity in China's ports has been expanded considerably. By the end of 2012, China's coastal ports had possessed more than 4100 productive berths with 6.49 billion tons of throughput capacity and 160 million TEUs of container handling capacity (Xu Xin and Hao Jun 2014). According to the analysis made by China Waterways Transport Research Institute of the Ministry of Transport, insufficient capacity is no longer the major concern of China's coastal ports (Zhang Baochen 2014).

In general, the future demand for ports in China will exhibit a steady growth. Yet it is unlikely for demand to resume the incessant and rapid growth in the foreseeable short term. First, the macro economy in China is facing downward pressure. Since 2012, the economic growth in China has slowed down to below 8 % from the previous 10 %. Second, the global economy of late is also not conducive to rapid growth in port demand. (1) The global economy is presently in the state of tepid growth, which not only implies that the new added value in world economy will shrink in the foreseeable future, but it will also trigger a series of contraction responses in international trade. (2) The re-industrialization strategy implemented by developed nations in Europe and the US following the 2008 Financial Crisis will to some degree change the international flow of raw materials, semi-manufactured and manufactured goods, thus slowing down the demand for China's ports. (3) The rise of cross-border e-commerce and the Free Trade Zones in China will envelop its major port regions, thus considerably reducing the volume upsurge caused by the previously irrational transport, while posing new challenges to the port service model.

When the port capacity has basically met the needs of social and economic development and when the future growth of port demand has slowed down, it is difficult for China to continue the port development model emphasizing scale expansion for the past decade or so. In the future, the port enterprises need to focus more on connotative development by improving the service quality, level and capability.

10.1.1.2 Less Government-Led and More Market-Driven Port Development

For years, port management system in China has been featured by the mingled functions of government and enterprises; port management and berth operations are intertwined, and port operations are in the relative state of monopoly led by the government. In light of this, the 3rd Plenary Session of the 18th CPC National Congress in 2013 resolved to deepen the reform in management system and improve the modern market system. It advocated to fully leverage the market's decisive role in resource allocation, and promoted resource allocation according to the market rules of maximizing the benefits and efficiency.

As required by the decree of deepening the comprehensive reform, China will significantly reduce the government's direct allocation of resources to the ports in the future. The government's main role in future port development is to enhance and optimize the public service, maintain market order, ensure fair competition, rectify market failure, and push forward innovation and sustainable development. In the future, port development will follow the market rules and pricing functions, so as to achieve maximization of benefits and optimization of efficiency through market competition.

10.1.2 Industrial Background of Port Transformation and Upgrading

10.1.2.1 Larger Ships Raising New Requirements for Port Development

Building larger ships has been a trend of global shipping industry, which has affected the ports in the following manner.

First, it has restructured the port hierarchy system. Large ships have highlighted the role and status of global and regional hub ports, forcing the neighboring ports to play the role of "feeder ports."

Second, it requires ports to improve their service level across-the-board in such aspects as hardware conditions, operating efficiency, operations management and connections upon landing. At the same time, it also requires port enterprises to upgrade or replace the loading/unloading machinery, improve the loading/unloading standards, thus better adapted to the berthing of large ships. Furthermore, it requires ports to have deeper seaways, berths of higher grade, stowage yard with larger capacity, and more convenient channels, thus ensuring the cargos to have smooth and unblocked flow from warehousing to distribution.

10.1.2.2 China's Ports Deficient in Service Functions and Risk Countercheck

Most of Chinese ports have simplistic operating model with their revenues relying chiefly on berth loading/unloading, warehousing and other traditional business. Except Xiamen and Tianjin, most port's income from port loading/unloading accounts for 60–90 % of its total revenue. However, the sources of income of international ports in the US and Hong Kong are quite diverse. Table 10.1 shows the makeup of operating revenue of typical Chinese and international port enterprises.

Table 10.1 reveals that the non-loading/unloading service in most Chinese ports (except for Xiamen and Tianjin) accounts for less than 50 % of port service. However, in developed nations, it accounts for 70 %. The functions in Chinese ports

Table 10.1 Makeup of operating revenue of typical Chinese and international port enterprises

Unit (%)

Port		LA	HK	Beihai	Xiamen	Tangshan	Tianjin	Shanghai	Dalian	Rizhao	Yantian	Ningbo
Port loading/unloading		23	29	60	18	61	33	59	73	85	81	62
Port logistics	Warehousing	4		1				17	27		2	38
	Tallying				4		1					
	Agency	7	34		10		3					
	Processing	9										
	Transport	28	28		9	2					17	
	Others				1	4						
Port service	Ship supply	17						24				
	Trade sales			39	46	27	53					
	Tow business				12		3					
	Port affairs management					6				9		
	Others	12	9				7			6		

Note: (1) Logistics processing includes packaging, testing and other logistics value-added service; (2) Transport includes ship transport, operation and management, and ground transport

Source: Liu Tiexin and Wang Haixia (2012), "Causes and tactics of Chinese port enterprise transformation and upgrading," Chinese Ports, 06:9–12

are relatively simplistic, such that port enterprises are weak to address the risk of rising cost. For instance, as the labor cost in China rises and investment in port security and environmental protection increases in recent years, port enterprises have run into difficulty in their operations.

10.1.3 Policy Background of Port Transformation and Upgrading

In recent years, to push forward transformation and upgrading of China's ports, to realize connotative development of ports and improve the competitiveness of ports, the Ministry of Transport has issued a number of policies designed to promote port transformation and upgrading. Some coastal provinces and cities have also issued like policies.

10.1.3.1 Policies at the National Level

In February 2009, the *Notification of the Ministry of Transport about Issuing Development Policies for Resource-saving and Environmentally-friendly Highway and Waterway Transport* was issued, which proposed that highway and waterway transport should adopt the development path featuring resource savings and environmental friendliness.

In June 2014, the Ministry of Transport issued the *Guiding Opinions on Pushing Forward Port Transformation and Upgrading*, which outlined the target, major tasks and upgrading paths of China's port transformation and upgrading.

The target of Chinese port transformation and upgrading is to basically establish, by 2020, modern port service system featuring high quality of benefits, strong hub functions, greenness, safety, intensive development, efficiency and convenience, so as to meet the needs of China's economic and social development. The leading role of port informatization will be more prominent, and the fundamental role of standards will be fully played. Basically, the port development will shift from reliance on resource input to reliance on technological progress, improvement of worker performance, and management innovations. It will also shift from mainly providing loading/unloading services to providing both loading/unloading services and modern port services, and shift from pursuing throughput increase to improvement on quality and benefits.

Major tasks of China's port transformation and upgrading encompass six aspects: (1) expanding service function and developing modern port business; (2) improving port transport system and pursuing the development of comprehensive port hub; (3) allocating port resources in a rational way and steering intensive development of ports; (4) enhancing technological and management innovations to push forward development of green ports; (5) strengthening port security management and bolstering port safety; (6) improving IT level of ports and promoting efficient and convenient port service.

Major paths of China's port transformation and upgrading include the following issues. (1) Reform and promote the transformation and upgrading. The reform focuses mainly on three aspects. First, reform the administrative hierarchy mechanism and establish the system of List of Rights. Second, enhance market-oriented reform and study the management model of Negative List. Third, reform of port charges towards market-driven and gradually liberate the competition of port charges. (2) Fully leverage the fundamental role of standardization. Enhance the formulation and revision of standards on energy conservation, emission reduction, environmental protection and safety management. Promote the application of China's port standards and participate in international standardization tasks. Encourage China to shift from a "Major Port Country" to a "Power Port Country." (3) Promote port transformation and upgrading via better planning. Safeguard the authority and solemnity of port planning, adjust and revise plans; make evaluation in accordance with Regulations on Port Planning and Management, initiate revision and adjustment procedure according to regulations, and only upon approval.

10.1.3.2 Policies at the Local Level

To promote port transformation and upgrading, local governments, based on their respective conditions, have also issued some policies and measures.

In 2011, the Shanghai Administration of Transport and Port issued the *Guiding Opinions on Building Shanghai Port into a Resource-saving and Environmentally-friendly Port,* which proposed the energy-consumption target and exhaust emission target in Shanghai Port by 2015 and 2020. In June 2014, Jiangsu Province issued the *Plan of Six Major Actions in Coastal Development,* which proposed the plan of improving the coastal port functions. In August 2014, Fujian Province issued the *Outline of Action to Speed Up Port Development,* which proposed the targets and eight major tasks in port transformation and upgrading in Fujian Province. In November 2014, Hebei Province issued the *Opinion of Hebei Provincial People's Government on Speeding Up Coastal Port Transformation and Upgrading and Providing Strong Support for Coordinated Development in Beijing, Tianjin and Hebei,* which proposed three strategic tasks in functional positioning, development goals and model of port construction.

10.2 Main Modes of China's Port Transformation and Upgrading

In recent years Chinese ports, by reflecting on their past development trail, have actively explored the issue of transformation and upgrading, and formulated some transformation and upgrading modes conducive to effectively expanding the port functions and boosting the quality and service level. These modes mainly include: (1) extending industry chain and implementing diversification strategy; (2)

implementing networked layout and enhancing network service capability; (3) linking with economic hinterland to provide impetus support for port transformation and upgrading; (4) elevating port management efficiency and service level through IT development; (5) constructing green ports to promote sustainable development of ports. These modes are discussed in the following sections in turn.

10.2.1 Extending Industry Chain and Implementing Diversification Strategy

10.2.1.1 Transformation from Traditional Port Logistics to Modern Logistics

Modern port has become an important component in international grand circulation system supporting world economy and international trade development, and a central link chaining global production, exchange, distribution and consumption. Modern logistics has progressively become a principal development direction of modern port; port logistics has gradually converted from disparate services of simple transport and warehousing to consumer-focused total-process logistics service.

Shanghai Port has proposed the "General Customs Clearance" program to actively develop new service fields of merchandise classification, packing, processing and distribution. It has also set up the Yangtze River Port Logistics Company to expand its business scope from simple terminal-handling tasks to providing cargo owners with door-to-door service. Tianjin Port has vigorously effected port-oriented total-process logistics program, and formed several logistics patterns such as auto logistics and cold chain logistics. Hebei Port Group has actively facilitated port construction for coal logistics, to build a complete industrial supply chain of coal consisting of trading, warehousing and distribution, circulation and processing, and high-end value-added logistics services for coal.

10.2.1.2 Emphasizing Construction of Bulk Commodity Transaction Market

Tianjin Port has established a bulk goods transaction market, built various market platforms covering auto, motor home, FMCG and bulk commodity, with fully-established functions of allocation, distribution, supervision and trading. The Beibuwan Port in Guangxi has fostered bulk commodity markets of coal, oil, iron ore, manganese ore, nickel ore, sulfur, heavy metal and grain. Hebei Port Group, relying on its ample resources, has constructed an electronic coal transaction platform. Dalian Port is now endeavoring to build an auto transaction platform and a grain transaction platform. Zhoushan Port in Zhejiang Province is striving to build an international bulk materials platform and an energy commodity transaction platform.

10.2.1.3 Developing Logistics Finance Business

Tianjin Port began to build a logistics finance platform in 2008, to engage in such business as commodity supervision, real-time commodity status control, and financial risk abatement, and thus establishing a smooth logistics finance system. In 2012 the Tianjin Port Bulk Commodities Transaction Company concluded a strategic cooperation agreement with Ningxia Tianyuan Manganese Industry Co., Ltd. for deepened cooperation in logistics finance and supply chain management, so as to build a "logistics finance chain."

10.2.1.4 Developing Port Real Estate

The *Guiding Opinions on Facilitating Port Transformation and Upgrading* issued by the Ministry of Transport in 2014 encourages port enterprises to develop port real estate and rationally utilize coastline resources. In recent years Shanghai Port and Tianjin Port have endeavored to develop port real estate. For example, Shanghai Port has directed the construction of commercial port real estate projects such as "Shanghai Long Beach," Huishan Plot and Haimenlu Plot.

10.2.1.5 Developing Ocean Liner Economy

Ocean liner industry, as an essential aspect of port transformation development and industry structure adjustment, has attracted much attention of all ports. By the end of 2013, 16 ports in China had experience receiving international ocean liners. Four home ports for international ocean liners have been put into operation in Shanghai, Xiamen, Tianjin and Sanya. Hong Kong, Dalian, Qingdao, Zhoushan and Yantai ports are making efforts to construct ocean liner terminals, and Zhuhai and Haikou ports are also planning the construction of ocean liner terminals (Yan Chenguang and Zhu Binjiao 2014).

10.2.2 Implementing Networked Layout and Enhancing Network Service Capability

Chinese ports and dock operators generally carry out strategic cooperation with domestic and foreign ports, shipping enterprises and logistics enterprises through investment in and operation of domestic and foreign docks, to provide customers with network-based service capability.

10.2.2.1 Building Domestic Port Network

(a) Building Domestic Port Network through Investment and Sharing

Shanghai Port has implemented the "Yangtze River Strategy" through dispensing capital, technology and management to cooperate with all regions along Yangtze River. In recent years, Shanghai Port has successively made investment in container terminals in five cities including Nanjing and Jiangyin, and in port groups in Wuhan and Jiujiang; it has realized holding of and sharing in four shipping companies, with shipping lines covering ports in provinces and cities such as Jiangsu, Zhejiang, Anhui, Jiangxi, Hubei, Hunan, Sichuan and Chongqing, and formed a complete "Yangtze River Economic Chain."

The Ningbo Port Group has deepened its cooperation with ports around the Yangtze River Delta, including Shanghai, Nanjing, Taicang and Lianyungang. It has constructed specialized bulk goods dock in Taicang and Nanjing to forge placement along Yangtze River, and build an "enclosed circle" of port economy.

(b) Establishing Regional Port Alliance

In 2011, seven port groups from Nanjing, Zhenjiang, Yangzhou, Taizhou, Jiangyin, Zhangjiagang and Nantong along the downstream of Yangtze River established the "Joint Meeting of the Seven Downstream Ports" and set up a regular meeting mechanism for high-level leaders from these regions, to jointly facilitate inter-port business cooperation and regional logistics cooperation.

In December 2014, ports and relevant enterprises in 16 cities along Yangtze River such as Yibin, Yichang, Wuhan, Hefei and Nanjing jointly launched the "Joint Meeting for Regional Cooperation of Port in Yangtze River Economic Zone," which helps to further invigorate cooperation of all ports and shipping enterprises in business, capital and regional logistics, so as to form a mutually beneficial development pattern of ports along Yangtze River.

(c) Government-led Regional Port Integration

Zhejiang, Fujian, Guangxi, Hebei and Chongqing, under respective government's leadership, have started internal port integration, to facilitate unified operations and management of port business and avoid redundant construction and vicious competition.

As far back as in December 2005, the Zhejiang Provincial government had integrated Ningbo Port and Zhoushan Port into one and set up Ningbo——Zhoushan Port Management Committee, to be chiefly responsible for the planning and management of Ningbo and Zhoushan ports, the compilation, reporting and synthesized release of statistical data of the ports, and the coordination of publicity and corresponding investment promotion. In September 2014, Ningbo Port Group and Zhoushan Port Group concluded the framework agreement for strategic cooperation, in order to actively conduct Ningbo—Zhoushan Port's development and construction with "capital as the tying belt and project as the carrier."

Table 10.2 Main events in China's port enterprises' layout of international network

Date	Port enterprise	Events in layout of international network
2008	COSCO Pacific Limited	Acquired operating right of Terminal 2 and Terminal 3 of the Port of Piraeus, Greece.
2011	China Merchants Holdings (International) Company Limited	Established JV Colombo International Container Terminal (CICT) with Sri Lanka Port Administration, winning a contract for 35-year construction and operation of southern container terminal of the Port of Colombo.
2013	China Merchants Holdings (International) Company Limited	Made investment of 400 million Euros to purchase 49 % of shares of Terminal Link Port Company wholly owned by CMA-CGM; Terminal Link holds equity of multiple docks in France, Morocco, Malta, US, Coate d'Ivoire, Belgium, China and Korea, etc.
2014	Qingdao Port	Concluded the Framework Agreement for Enhancing Port Strategic Cooperation between Qingdao Port Group and Dubai Ports World with Dubai Ports World.
2014	Lianyungang Port	Concluded the JV agreement for "Sino-Kazakhstan Cooperative Lianyungang Logistics Yard Project" with Kazakhstan Railway Transport Company, to establish a JV company to construct and operate logistics yard project in Lianyungang Port.
2015	Shanghai Port	Won the bid for terminal operating right of Haifa New Port, Israel from 2021 to 2046.

10.2.2.2 Layout of International Network

In recent years, China's ports have started implementing the international development strategy and gradually laid out international network through investing, share holding, franchising, and strategic alliance and cooperation. Table 10.2 illustrates the details.

10.2.3 Joint Action with Economic Hinterland to Provide Impetus for Port Transformation and Upgrading

Port hinterland and port are interdependent and interactive. The more advanced the hinterland economy is, the higher the transport demand for port will be, which tends to drive port size expansion and service upgrade. Port development can create conditions for hinterland economic development and promote further expansion of port hinterland's scope.

10.2.3.1 Enhancing Economic Ties with Hinterland Through Dry Port

The inland logistics network with dry ports as nodes has become a notable tie connecting ports and inland areas. Tianjin Port has constructed 23 dry ports and five regional marketing centers in interior hinterland, with dry port's business scope covering import and export Customs declaration, inspection, goods transport, warehousing, and container management (Mao 2014). This regional Customs clearance mode could accomplish the "One-pass Customs Declaration" and immediate release of cargos upon arrival at port. Tianjin Port also offered "Online Dry Port" in April 2014, to integrate and lift its functions and service level through online and offline collaborative service, further enhancing economic ties between Tianjin Port and its hinterland area.

10.2.3.2 Enhancing Economic Ties with Hinterland Through Sea-railway Coordinated Transport

In recent years, Ningbo Port has focused on the construction of two sea-railway coordinated transport lines—the "Thousand-kilometer Zhejiang-Jiangxi Railway" and the "Ten-Thousand-kilometer Ningbo—Xinjiang—Europe Railway." The Ningbo-Jiangxi special train of sea-railway coordinated transport has achieved "daily availability," and the Ningbo—Xinjiang special train of sea-railway coordinated transport has extended port hinterland to the Silk Road Economic Belt.

Lianyungang Port has kept close cooperation with inland areas, established strategic cooperation relations with ports in Henan, Shaanxi, Gansu and Xinjiang, and opened the "Five-scheduled" regular container trains to and from Houma, Taojiazhai, Wuxi and Alashankou.

10.2.3.3 Enhancing Economic Ties with Hinterland Through Water-Water Coordinated Transport

In recent years Shanghai Port has facilitated the "Yangtze River Development Strategy" all along, mainly through the following measures: first, actively pushing forward the implementation of standardization of river-sea coordinated transport; second, uniting ports along Yangtze River to accelerate information innovation and enhance information connection between originating port and destination port; third, encouraging shipping companies to conduct business along Yangtze River through river-sea direct links. Shanghai Port has also focused on developing water-water transshipping lines from Anji Chuanda Inland Port, Jiaxing Inland Port and Dushan Port to Shanghai Port, effectively enhanced the radiation strength of Shanghai Port to North Zhejiang and South Jiangsu.

10.2.4 Lifting Port Management Efficiency and Service Level Through Informatization Development

Informatization construction of port enterprises is the fundamental requisite in developing modern shipping industry, and a main avenue to improve port management efficiency and service level. Port informatization development mainly includes

Table 10.3 Construction of management information platform in some port enterprises

Port	Main contents of management information platform construction	Effect achieved
Shanghai Port	In Jan, 2014, the "Ship Dispatching Platform" was officially put into operation, to realize TOS (Terminal Operating System) real-time data transmission, supported with on-site video, to enable port enterprises to control real-time information of ships by terminal whenever necessary.	The platform helped achieve the standardization of multiple codes of shipping information, ships, fares, customers, terminal, container and efficiency, and consistency of statistical form.
Dalian Port	The "TOP+ Intelligent" container terminal operation system. Application of Internet of Things and Internet technology allows the container terminal to operate completely by electronic instruction.	Reduced number of business personnel on the spot, decreased safety risk and labor cost, lifted data collecting and processing timeliness.
Tianjin Port	Substituting RF card for packing list, all information in container packing list such as ship name, voyage and B/L No, etc. can be recorded into the RF card.	Created possibility for cancelling gate-in packing list and streamlining gate-in operation flow.

construction of port management information platform, construction of electronic port, and informatization development at logistics centers.

10.2.4.1 Construction of Port Management Information Platform

Construction of port management information platform can boost work efficiency, provide port operators, ship owners and cargo owners with more complete and speedier service, and lift port administration's performance level. Construction of management information platform in some port enterprises are shown in Table 10.3.

10.2.4.2 Construction of Electronic Port

In June 2014, Shanghai Port conducted the "Single Window" pilot operation in Shanghai Bonded Port. The main functions of "Single Window" consist of six modules, i.e., goods declaration, transport vehicle declaration, import and export permit, enterprise qualification management, payment settlement, and information inquiry. In 2015, the "Single Window" practice was popularized in coastal ports. Currently, coastal ports in major trade provinces such as those in Tianjin, Guangdong, Fujian, Zhejiang, Jiangsu and Liaoning are gaining momentum in establishing the "Single Window" operation.

10.2.4.3 Informatization Development at Logistics Centers

Informatization development at logistics centers aims to establish modern logistics, by closely knitting cargo owners, customers, domestic and foreign trade companies, warehousing companies, transport companies, production and supply and sale units, so as to attain common development, construction, management, and operations. The ultimate goal is to realize the strategic pattern of "plant, road, port and transport in one fell swoop, and production, supply, sale and trade as one unit."

Qingdao Port's Seaway Network is a logistics service platform built on the basis of e-commerce and network public platform, by integrating the logistics resources of industries and the customer resources of trading enterprises in Qingdao area. The Seaway Network makes use of three major blocks of platforms, namely the shipping schedule and freight platform, the cabin booking service platform, and the media platform, to achieve convenient information retrieval, effective cabin booking and building of the city's largest information bank.

Hebei Port Group has taken advantage of technologies in information sharing, data mining, cloud computing and virtualization to attain deep integration of information technology with the port's traditional coal business.

10.2.5 Construction of Green Port to Promote Sustainable Development

In recent years, Chinese port development has paid mounting attentions to the negative impact of ports on regional environment, and gradually sped up the construction of new-type of ports featuring low energy consumption, low pollution, and high efficiency.

10.2.5.1 Advocating the Adoption of "On-shore Power" Mode

Since 2010, the Ministry of Transport has successively launched the demonstration of using on-shore power for ships docking at seven berths of the container terminals in Shanghai Port, Lianyungang Port and Shekou Port. In 2012, the Ministry of Transport instituted and published the technical standards, rules for safe operation, and design norms for using on-shore power for docking ships; it also decided to reward projects already using shore power by compensating 20 % of the project construction cost.

Owing to the attractive support of above policy, some domestic ports intensified their investment in on-shore power projects. Currently, of the ports in China, Shekou, Cangzhou, Ningbo, Tianjin, Shanghai, Lianyungang, Shenzhen Yantian and Huanghua ports have had "Shore Power for Vessels" project in operation.

Table 10.4 Some practices of green ecological port construction in China

Port	Typical measures	Goal and effect
Qingdao Port	"Optimal Route" for container trucks	Saving energy of 1344 tons of standard coal annually.
Ningbo Port	Container terminal "One tractor Two trailers" transport mode	Saving fuel by 44 % from heavy container transition, by 40 % for empty container transition, and energy by 37.25 % overall.
Lianyungang	Comprehensive technology for energy saving and emission reduction in 300 k-ton ore terminal of Xinsu Port	Annually saving electric energy by 2.323 million kW, reducing carbon dioxide emission by 3476 tons and saving fuel by 379 tons, equivalent to standard coal of 1490.7 tons.
Tianjin Port	"Zero Idle Variable-speed Operation" of the specialized berth for coke	Lifting production efficiency by 29.9 % and saving electricity by 4.056 million kW per year, equivalent to standard coal of 1638.5 tons.
Rizhao Port	Comprehensive energy saving technology for ore	Saving electricity by about 765,000 kW (costing nearly 800,000 RMB), namely saving energy equivalent to 254.7 tons of standard coal.
Chongqing Port	Popularization of new-model 325 TEU energy-saving container ship	Annual energy savings equivalent to standard coal of 4330 tons, reducing carbon dioxide emission of 10,640 tons, NO_x emission of 67.55 tons, SO_x emission of 71.45 tons, and smoke dust emission of 41.57 tons.

Source: Wang and Wang (2013) Current situations of green ecological port construction in China and the countermeasures. World Shipping, 36(3):9–12

10.2.5.2 Encouraging Ports to Use LNG

In October 2013, the *Guiding Opinion of the Ministry of Transport on Facilitating Application of LNG in Water Transport Industry* was issued. In September 2014, the General Office of the Ministry of Transport issued the *Circular on the Implementation Scheme for Pilot and Demonstration of Application of LNG in Water Transport Industry*, and offered seven pilot projects, six demonstration projects and three demonstration zone projects.

10.2.5.3 Other Ecological Port Construction Practices

Besides generally-adopted shore power technology and use of LNG in ports, all ports have carried out numerous practices in green port construction, based on each port's own characteristics as detailed in Table 10.4. For example, Qingdao Port implemented the project of optimal routing for container trucks, capable of saving energy of 1,344 tons of standard coal annually.

10.3 Outlook for Transformation and Upgrading of China's Ports

Port transformation and upgrading is a long process. To reach the goal, China's ports, on the one hand, must match their efforts with national major development strategy to better serve the national economic and social development; on the other hand, they should be guided by technology progress. Rapid development of new modes such as "Internet +," cloud computing, big data, Internet of Things and e-commerce, and other new technologies will bring about significant influence on port transformation and upgrading.

10.3.1 Greater Importance Attributed to Service Capability of Ports' Comprehensive Supply Chain

The principal direction of China's ports' transformation and upgrading lies in concertedly developing modern port logistics based on supply chain management, and promoting transformation of traditional port handling and transport enterprises to modern port logistics enterprises.

Increasing number of ports has planned transformation towards comprehensive logistics supplier. For example, according to the national planning, Shanghai Port will develop the new business modes of auto logistics, regional dispersing center and cold chain warehousing, so as to accelerate the transformation from a singular server to a comprehensive logistics supplier. Tianjin Port has proposed to make logistics service functions broader and stronger, so as to better provide services of consolidation and transit, transport warehousing, processing distribution and logistics finance; by so doing it will speed up the progress toward an international logistics center and resources allocation center. Hebei Port Group will rely on port-centered collection/dispersion/distribution system, and analyze the needs of buyers such as power plant, to carry out mixing and processing of coal from sellers and realize added value in logistics supply chain.

10.3.2 Ports' Service Functions Will Be Enhanced Further

Full-scoped implementation of the Pilot Free Trade Zone policy in China in recent two years will favorably promote further enhancement of ports' service functions.

10.3.2.1 "Pilot Free Trade Zone" Policy Will Greatly Improve Domestic Ports' International Transit Functions

Currently, China's ports are basically of hinterland-type, having minimal international transit volume. Pilot Free Trade Zone's open environment will attract exports from Southeast Asia to transit at Shanghai, Shenzhen, Fuzhou and Tianjin ports, and allow non-five-star flagged ships to pick up and drop off goods at coastal ports. This will change the former ways of domestic goods going through Busan for international transit to going through domestic ports for transit. If relevant policies are relaxed to allow foreign-funded companies to pick up and drop off at coastal ports in the future, domestic ports' international transit capability will be further enhanced.

10.3.2.2 "Pilot Free Trade Zone" Policy Drives Ports to Change from Logistics Transport Feeder Port to Logistics Transaction Integration Port

A series of policies issued by the Pilot Free Trade Zone will help ports to perfect spot trading platform, expand futures transaction functions, accelerate the development of "logistics + finance," "logistics + trade," and "logistics + e-commerce" businesses; this will drive ports to change from traditional logistics transport feeder port to logistics transaction integration port.

First, Pilot Free Trade Zone's policies for bulk commodity transaction and port periphery commerce development provide favorable conditions for port enterprises to engage in commerce and trade businesses. The *Overall Plan for China (Shanghai) Pilot Free Trade Zone* proposes "exploring establishment of international bulk commodity transaction and resource allocation platform, and conducting international trade of energy products, basic industrial materials and bulk agricultural products in the pilot zone." This stipulation creates auspicious conditions for port enterprises to engage in port periphery commerce and trading businesses. The *Overall Plan for China (Tianjin) Pilot Free Trade Zone* permits lawfully conducting bulk commodity spot trading, explores establishment of foreign exchange management and Customs supervision systems adapted to international bulk commodity transaction.

Second, the Pilot Free Trade Zone affords the policy support for the development of logistics finance. The *Overall Plan for China (Shanghai) Pilot Free Trade Zone* suggests "expansion and perfection of bonded settlement of futures pilot project and expansion of functions of warehouse receipt financing," thus providing development space for port enterprises to expand logistics finance service.

10.3.3 Ports' Global Network Layout Will Be Further Accelerated

Modern port, as a link in supply chain, emphasizes interaction with other ports. Terminal operators' global networked layout through expansion and merger is a prominent characteristic in modern port development.

In 2015, the National Development and Reform Commission, the Ministry of Foreign Affairs, and the Ministry of Commerce jointly released the *Vision and Actions on Jointly Building the Silk Road Economic Belt and 21st Century Maritime Silk Road*, proposing to "direct the construction of ports infrastructure, smooth the land-water coordinated transport passage, facilitate the port cooperation, increase the marine shipping lines and flights, and enhance the marine logistics information cooperation."

China's port enterprises have ample experience in infrastructure construction and operations, whereas nations in the Southeast Asia and South Asia along the 21st Century Maritime Silk Road have strong demand for construction of large ports. Therefore, Chinese port enterprises face outstanding "going abroad" opportunities. It can be expected that under the "One Belt and One Road" strategy, Chinese ports will further speed up the pace of "going abroad" in the future to participate in major foreign ports' construction and operation, and be integrated into the global shipping system.

10.3.4 Inter-regional Port Integration Will Be Further Intensified

China has put forward the collaborative, balanced and highlighted regional development strategy, underpinned by the Beijing-Tianjin-Hebei Coordinated Development and the Yangtze River Economic Belt projects. The regional coordinated development strategy poses certain new requirements for inter-regional integration of ports, and the State has also mapped out relevant policies to invigorate inter-regional integration of ports.

10.3.4.1 Horizontal Integration of Regional Ports Will Be Enhanced

The trend of assets-tied horizontal integration of regional port groups has emerged. For example, in August 2014, the Bohai, Tianjin and Hebei Port Investment and Development Co., Ltd. was officially established with joint investment by Tianjin Port Group and Hebei Port Group. In the future, the company will follow the marketization principle to further optimize resource allocation of regional ports, achieve complementary location advantage, and foster the crucial role of ports in propelling the development of Beijing, Tianjin and Hebei. Ports in the Yangtze River Delta region will form a strategic alliance centering on the goal of establishing Shanghai international shipping center in the future, and to construct the port cluster of Shanghai international shipping center through mutual investment, share holding and joint construction of dock berths.

10.3.4.2 Division by Specialization and Functional Postures of Regional Ports Will Be Fortified

In the future, competition of ports will be mainly embodied in the manner of competition among port clusters. Against the backdrop of the regional collaborative development strategy, specialized and rational division among regional ports under policy support may yield the breakthrough in raising the competitiveness of port clusters. At present, the trend of specialized division among regional ports has emerged to some extent. For example, according to the plan of Shanghai International Port (Group), Waigaoqiao will no longer accept the business of transiting Yangtze River upstream containers to Yangshan Port; the transit business will be completely transferred and centralized to Taicang Port (Wang Minjie 2014).

10.3.5 Ports Will Be Upgraded to the "Internet + " Phase

The 2015 Government Work Report firstly proposes the "Internet+" action plan to incorporate Internet into the national strategy. Implementation of the "Internet+" strategy will compel ports to apply Internet thinking, Internet of Things technology, big data technology, cloud computing technology, geographic information system (GIS) and other information technologies to port production. In the future, "Internet + port" informatization trend will be manifest.

"Internet + port" upgrades port from a spatial node in the ocean shipping logistics chain to an information service node; it helps in port integration and the extension of port logistics service industrial chain, and guides port enterprises to alter from production orientation to service orientation. "Internet+" will play a role in port upgrading in two aspects. First, in terms of port's internal communication, it will extend Internet to sensor network, to form the port Internet of Things, achieve automatic data collection and processing, real-time online tracking, monitoring, statistics gathering, and inquiry. Furthermore, by integrating terminal information resources and via port big data platform and cloud computing platform for terminal dispatching operation, it will attain standardized and intelligent terminal operations. Second, port can better explore and meet customers' demands for service. On the one hand, it will be easier for port operators to perceive changes in demand of the service objects such as cargo owners and ship companies, so as to make port services closer to market demand. On the other hand, the key customer-oriented business mode in port service will be overturned; ports can rely on Internet platform to provide small/med-sized customers with customized service, so as to stimulate their demand for port service.

10.4 Summary

This chapter mainly presents the background, the modes, and the future development trends of China's port transformation and upgrading. Due to changes in international and domestic economic conditions, growth in port demand will slow down in the future, and China's port development mode focusing on scale expansion will be hard to continue. With the intent for comprehensively deepening the reform expounded in the Third Plenary Session of the 18th Central Committee of the CPC, future development of China's ports will rely more on the sway of the market. The development trend of larger ships has also raised new requirements for port development. Against such macro and industrial background, China's Central Government and local governments in coastal regions have released policies to promote port transformation and upgrading. Through active exploration, China's ports have formed five transformation and upgrading modes: (1) extending industry chain and implementing diversified development strategy; (2) implementing networked layout, enhancing network service capability; (3) taking joint action with economic hinterland to provide impetus for port transformation and upgrading; (4) lifting port management efficiency and service level through informatization; (5) building green port to promote sustainable development of ports. China has implemented a series of major development strategies: the "Pilot Free Trade Zone" policy, the "One Belt and One Road" strategy, the "Beijing, Tianjin and Hebei Coordinated Development" strategy, the "Yangtze River Economic Belt Development" strategy, and the "Internet +" strategy. To better implement China's major strategy and meet demands of economic and social development, ports' transformation and upgrading will exhibit the following attributes. (1) Greater emphasis will be placed on the service capability of port's comprehensive supply chain. (2) Ports' service functions will be further lifted. (3) Ports' pace in global network layout will be accelerated. (4) Regional integration of ports will be further enhanced. (5) Ports will be upgraded toward "Internet + ."

References

Liu T, Wang H (2012) Exploration of the incentive and strategy for transformation and upgrade of Chinese Port Enterprises. China Ports 6:9–12

Mao, Z (2014) Tianjin port constructed 23 inland "Dry Ports", with 70 % of the throughput from inland, Netease, http://money.163.com/14/0512/15/9S2BAMJC00254TI5.html. Accessed 12 May 2014

Wang, M (2014) Yangtze river downstream ports approached cooperation-competition, Reform of customsclearanceintegrationlaunched.ifeng.http://finance.ifeng.com/a/20141106/13252873_0.shtml. Accessed 6 Nov 2014
Wang S, Wang Y (2013) Current situations of green ecological port construction in China and the countermeasures. World Shipp 36(3):9–12
Xu X, Hao J (2014) Thinking about deepening reform of management system of coastal ports of China. China Ports 2:1–4
Yan C, Zhu B (2014) Comprehensive evaluation on and development proposal for home port for ocean liners in China. Communication Shipping 5:52–58
Zhang B (2014) Government role in transformation and upgrading of Chinese port. Chinese Cadres Tribune 7:57–60

Chapter 11
Service Mode and Development Trend of the "Last-Mile Delivery" of E-commerce Logistics

Zhi-lun Jiao

The rapid development of e-commerce is pushing the profound transformation of China's traditional operations mode of logistics; wherein the logistics service of "Last-mile" carrying and delivery, which faces the end consumers, is continually innovated, showing a great development potential. Under the support of the rapid development of e-commerce, the "Last-mile Delivery" service mode of China's e-commerce logistics is increasingly diversified, and has become an important component of urban community service and the citizens' livelihood. This chapter addresses the development background, the characteristics, the service modes, the development status, the existing problems and development trend, and other relevant topics of the "Last-mile Delivery" service of China's e-commerce logistics. Section 11.1 presents the concept of the "Last-mile Delivery" service of e-commerce logistics and its background factors of drawing much social attention; meanwhile, it discusses the chief characteristics of the "Last-mile Delivery" service of e-commerce logistics. Section 11.2 introduces the major service modes of the "Last-mile Delivery" of e-commerce logistics from four aspects: innovation services in express enterprises, innovation services in platforms, professional services of specific commodities, and professional services in specific regions. Section 11.3 describes the development status of the "Last-mile Delivery" service of China's e-commerce logistics and its existing problems, covering the continually expanded scale, the intensive innovation of service modes, the gradually formed market segments, and the constantly improved professional services. Section 11.4 presents the development trend of the industry, emphasizing issues on cross-border cooperation, service integration, professionalization, localization, adoption of new technology, energy conservation, and environmental protection.

Z.-l. Jiao (✉)
Logistics Research Center, Nankai University, Tianjin, People's Republic of China
e-mail: jiaozhilun2002@hotmail.com

L. Wang et al. (eds.), *Contemporary Logistics in China*, Current Chinese Economic Report Series, DOI 10.1007/978-981-10-1052-1_11

11.1 Development Background and Characteristics of the Last-Mile Delivery Service of E-commerce Logistics

With the rapid development of Internet economy, online integration is gradually realized by the social business system in three virtual processes, combining business flow, fund flow and information flow; logistics has developed into a relatively independent physical flow and taken on the development characteristics of professionalization and segmentation. Because e-commerce esteems the end-users' consumption experience, the delivery process of last-mile delivery at the terminal point of e-commerce logistics has emerged as a hot topic as well as a snag in China's current e-commerce logistics and urban logistics development

11.1.1 Concept of the Last-Mile Delivery Service of E-commerce Logistics

Logistics service includes a series of processes, wherein the last-mile delivery logistics service generally means the terminal distribution and delivery links of bringing the goods to the final customers, according to their different requirements. The complete process involves transportation, loading and unloading, warehousing and distribution, distribution processing, city delivery and other operations links.

In China, the last-mile delivery service usually has specific meaning due to the diverse logistics segmentations. For instance, "Last-mile Delivery" in highway freight field means the urban delivery phase after the completion of the transportation via trunk line and special line; the "Last-mile Delivery" of express logistics mostly means the community delivery phase of city ground delivery; the "Last-mile Delivery" of railway freight mostly means the final delivery phase from the railway station to the factories and the delivery destinations designated by the customer after the completion of trunk railway transportation. Based on the character of the final customers, the "Last-mile Delivery" is further divided into terminal delivery service to the commercial enterprises (2B) or terminal delivery service to the final customers (2C).

The "Last-mile Delivery" service of e-commerce logistics treated in this chapter means the process of achieving the distribution and delivery of the packages to the terminals, such as urban community and self-pickup depot, according to the requirements of the final customers, after online retail packages have gone through a series of logistics service processes. From the final customers' perspectives, the "Last-mile Delivery" service of e-commerce logistics mentioned in this chapter means the distribution and delivery service aimed at final customers (2C), excluding city delivery service aimed at enterprises or retail terminals (2B).

In the whole service process of e-commerce logistics, the "Last-mile Delivery" is the only link having face-to-face contact with the final customers. Therefore, the

"Last-mile Delivery" service is an important component for customers' experience in e-commerce logistics, and the direct contact point and service window with consumers in the whole supply chain of e-commerce.

11.1.2 Development Background of the Last-Mile Delivery Service of E-commerce Logistics

The development of the last-mile delivery service of e-commerce logistics is a logistics service mode in which logistics service network has gradually extended downward towards cities and communities based on the rapid growth of internet economy and e-commerce. As e-commerce consumption economy is developing towards personalization in products and services, the last-mile delivery service of e-commerce logistics has attracted wide-spread attention of the public.

11.1.2.1 Transformation and Upgrading of the Express Industry

The development of e-commerce has driven the transformation and upgrading of China's express industry. Competition among express enterprises began with price-only competition but eventually transformed to the competition in network coverage and service level. Terminal network in cities for serving the last-mile delivery is not only the direction of logistics network expansion, but also one of the key points for express enterprises to improve their competiveness. The layout of last-mile delivery of an express enterprise not only concerns the service quality, logistics effectiveness and cost of its logistics service, but also is closely related to its value-added services, including collection on delivery and home pick-up. Hence, it has become an important pillar for an enterprise to improve its competitiveness.

11.1.2.2 Gradually Personalized Consumption Demand

The target customers for the last-mile delivery are mainly young people who enjoy online shopping. They have become accustomed to shopping online; their consumption demand is more personalized and their requirement towards logistics service is more diversified. With the gradual elevation of Chinese people's living standard, different online consumer groups, especially young consumers, require more personalization in delivery time, location, and modes; their requirement for the quality of delivery service has also elevated. Continual expansion of the types of online shopping products – fresh foods, perishable foods, large home furnishings, small lot/multiple batches of instant consumption goods, require different terminal deliveries, and result in constantly improved service level of the last-mile delivery enterprises.

11.1.2.3 E-commerce Logistics Channel Extending Downward

Since mid-'90s, with the gradual popularization of the Internet in China, delivery service of e-commerce logistics has also developed rather swiftly. Driven by the booming network retail platforms and various C2C, B2C e-commerce, home delivery service has become the crux of the development of e-commerce logistics; meanwhile, e-commerce logistics enterprises also began to pay attention to the delivery service of the last-mile delivery in allusion to city communities. Propelled by the rapid growth of mobile e-commerce and the O2O[1] business mode, development of community and localization market of e-commerce has deepened; service network of e-commerce logistics penetrated into residence centers. Logistics service network of the last-mile delivery sprawling to areas such as highly congregated city communities and relatively developed rural areas has taken form.

11.1.2.4 "Storehouse Eruption" of Express Service during Sales Peak

The volatility of online shopping demand creates a tremendous challenge to the last-mile delivery service, and especially, under the promotion on large shopping platforms during the "Double 11" and "Double 12"[2] shopping spree. The upsurge of online shopping orders within a short time caused temporary capacity shortage and the risk of "storehouse eruption" of parcels to the last-mile delivery service. In a single day of "Double 11" in 2014, Taobao.com and Tmall.com platforms of the Alibaba Group accrued a total trade volume of 57.1 billion RMB, generating 278 million packages for delivery. Promotion activities of these large-scale e-commerce shopping festivals have brought enormous pressure to the express network. In view of this phenomenon, express enterprises must improve their response in peak season by improving the last-mile delivery service, reduce the distribution pressure of packages in warehouses at all levels, and enhance the flexibility of their logistics service.

---

[1] "O2O," an abbreviation of "online to offline," means the integrative development between the online business mode of Internet and the offline business mode of stores.

[2] "Double 11" and "Double12" are large-scale promotion activities conducted on the special days of November 11 and December 12, by notable e-commerce websites typified by Tmall.com, Taobao.com and JD.com to boost sales. Since 2009, these festivals have become China's largest Internet commercial activities. The Alibaba Group Holding Co., Ltd. applied to the State Trademark Bureau for the trademark registration of "Double 11" in November 1, 2011, and obtained the exclusive right of the trademark in December 28, 2012.

11.1.3 Major Characteristics of the Last-Mile Delivery Service of E-commerce Logistics

Compared with the general logistics distribution and delivery process, the last-mile delivery of e-commerce logistics has its distinguishing characteristics. The basic cause for all these characteristics lies in that the express enterprises need to pay much more attention to the shopping experience of the end consumers, namely, meeting the consumers' individual demands on timing, services and convenience.

11.1.3.1 Highly Personal Demands

The last-mile delivery service of e-commerce logistics is mainly aimed at the end consumers; logistics distribution and delivery requirements of the consumers are dynamic and diverse; different consumers vary in their service requirements. For example, for student customers at school, distribution and delivery strategies must be arranged based on their school hours, but for residents in city communities home delivery time must be arranged according to various schedules of their daily activities. For another example, some consumers wish to have the packages delivered to their residence on upper floors of the building with no elevators, while other consumers wish to have the packages delivered to the reception office of their workplace for security reasons. In addition, some customers need to have the packages delivered as soon as possible, while other customers wish the packages delivered at specific point in time. Due to these different individual requirements, express logistics enterprises of e-commerce need to layout their distribution network and plan the delivery routes according to the dynamics of varied consumer requirements so as to provide the personalized and differentiated last-mile delivery service.

11.1.3.2 Imbalance and Bi-Directionality

Imbalance in end distribution and delivery services of e-commerce logistics is embodied in that of time and space. Imbalance in time means that the demand for service has slack season and peak season. For example, some e-commerce promotion activities would boost the demand for express service at certain times by a large margin. Imbalance in space means that the logistics service demand for last-mile delivery varies greatly among different regions and different communities. For example, last-mile delivery service between urban and rural areas varies markedly. For shopping malls, business districts and residential communities great differences exist in terms of road passage, security condition and delivery requirement. Due to the demands on return and exchange goods for online shopping, the last-mile delivery service of e-commerce logistics also has the bi-directional characteristics of intertwined forward and backward logistics. In addition, this bi-directional logistics also has an imbalance, ie, ordinary forward deliveries to residential communities are generally greater than the backward logistics of home pick-ups.

11.1.3.3 Ample Room for Development of Value-added Services

As the only interface which provides contact with the end consumers, the last-mile delivery service could possibly be the main carrier for e-commerce and express enterprises to expand their online and offline full-channel services. Thus it has ample room for the expansion of various value-added services and additional services. The most common value-added service modes include: delivery at specific time and to specific site, collection on delivery, reception of package, temporary storage, return and exchange, home pick-up, delivery of goods with choice at receiving site, evening delivery, providing instruction on usage of goods, try-on/try-out, installation and debugging, etc. The most common value-added community services include group purchasing for the community, selling consigned goods, self-help bank, self-service payment of expense bills, tools leasing, and equipment installation/maintenance. Theoretically, the community last-mile delivery service could become a convenient comprehensive service center in populated areas, and has room for expanding into an all-around life service business.

11.2 Major Service Modes in the Last-Mile Delivery Service of E-commerce Logistics

At present, China's e-commerce industry is in a co-mingling and stacking phase, encompassing desktop e-commerce, mobile e-commerce, cross-border e-commerce and online and offline coordinated e-commerce. E-commerce logistics, as it complement, is also constantly innovated in areas of order processing, warehouse distribution and city distribution. In the area of last-mile delivery service of e-commerce logistics, many enterprises are actively exploring innovation in service modes, and have formed four major service modes. These modes are termed innovative service mode in express enterprises, innovative service mode in platforms, professional service mode of specific products, and professional service mode of specific locales.

11.2.1 Innovative Service Mode in Express Enterprises

Express enterprises are the major service providers of the traditional e-commerce logistics, and the earlier last-mile delivery service was relatively simple; it makes the terminal transport in the form of "city warehouse + transport to district," and completes the delivery in the form of "telephone contact + face to face delivery."

Table 11.1 General time-sensitive express products and value-added service products

Service products of express company	Time-sensitive express products	8-h intra-city delivery
		12-h next morning delivery
		24-h next day delivery
		36-h third day delivery
	Value-added services	Payment collection on delivery
		Fee collection for customers
		Goods picked-up
		Returning of signed bill
		Electronic receipt
		Insurance of express item
		Integrated warehousing/ delivery service

Source: Compiled from data contained in the websites of various express enterprises

Wherein, the operation of the city warehouse includes own-operated website or franchised "ground delivery" company.[3] With the growth in business volume, the last-mile delivery service providers in express enterprises have often innovated their service modes, and constantly improved their service efficiency.

11.2.1.1 Differentiated Mode of Service Products

Differentiated mode of service products allows consumers to choose delivery time, location, and form (such as face-to-face delivery, designated receipt, self pick-up) on their own as needed. Service providers have formed several relatively standard service products according to the consumers' preferences, so as to reflect the variations in the last-mile delivery services and set charges according to the differentiation. This is the differentiation mode of terminal distribution adopted by the express enterprises since earlier days. For example, the current express companies, including YTO, ZTO and STO, while providing standard express services, also provide time-sensitive express products, including the 8-h city delivery, 12-h next morning delivery, 24-h next day delivery, and 36-h third day delivery. They also provide integrated value-added full services including payment collection on delivery, fee collection for customers, goods pick-up, returning of signed bill, electronic receipt, insurance of express item, and warehousing/delivery. This differentiation mode of service product not only realizes personalized service and improves consumers' experience, but can also help understand the consumer's real needs, and accumulate

[3] "Ground delivery" company is a regional express company serving the terminal distribution of city communities; its terminal dispersing and delivery network is established generally through direct-operation or franchising in larger cities with large demand of express packages. The courier in charge of a designated district area makes the home delivery. The general business of the "ground delivery" company includes community dispersal, intra-city transfer and face-to-face delivery, etc.

the data for community services, which has practical significance in improving service efficiency, and reasonably controlling service cost. Table 11.1 shows the time-sensitive express products and value-added service products provided by China's express enterprises.

11.2.1.2 Pre-Service Mode Based on Big Data

The accumulation of big-data of logistics demand is a major characteristic which differentiates e-commerce express logistics from traditional logistics. In the last-mile delivery service, the prior-stage operations of logistics are planned and implemented in advance according to the results predicted from the previous big data demand; commodities are deployed to regional warehouses. The last-mile delivery service is transformed to the "Only-mile Delivery" after the customers place their orders on the online retail platform. Currently, this pre-service mode has become the way express enterprises conduct the terminal express service in order to cope with the peak demand period.

For example, during the "Double 11" shopping festival in 2004, the TMall.com platform merchants took advantage of the warehouse network of Cainiao Logistics Co. to help its affiliated merchants to pre-stock their merchandises. Based on factors such as merchandise types, historic transaction data, and efficiency in logistics routes, Cainiao suggested a stocking volume for each commodity, which was aligned with the merchant's predicted value. The commodity of pre-determined quantity was put into the regional distributed warehouse by the express enterprise in advance. Upon receiving an order from the customer, the express company dispersed the goods directly from the last-mile delivery warehouse. This service became an important content for TMall.com to carry out the "Same day delivery" service on the "Double 11" day for consumers in many cities. As another example, even though the APEC meeting was held in Beijing on the day of the "Double 11" shopping festival when the city's traffic was under strict control, yet 95 % of the consumers' orders still received the same day delivery service because JD Logistics made a prediction of the demand basing on the big data and deployed the commodities in advance.

11.2.1.3 The Crowd-Sourcing mode

Since the last-mile delivery of city express entails huge amount of human resource and transportation resource, some express companies attempt to employ the "crowd-funding"[4] idea of the Internet to reduce the social operational cost of the last-mile

[4] "Crowd-funding," evolved in the area of Internet finance, refers to a behavior involving the initiator, the investors and the platform, which raises funds from the masses to support and initiate some investment project. Generally, Crowd-funding links the sponsors and the proposers through the network platform. At present, it is widely adopted in projects such as post-disaster reconstruction,

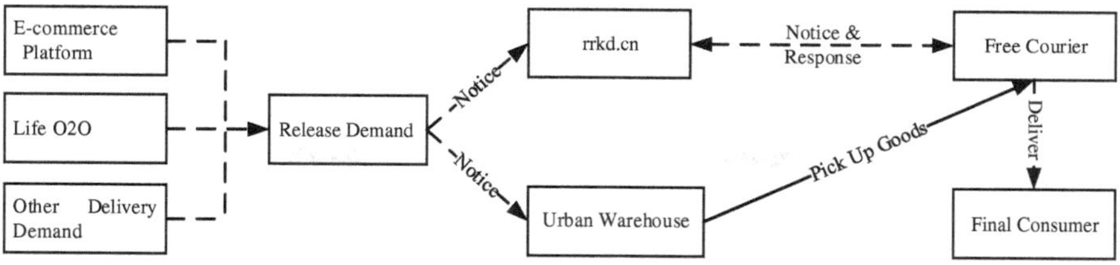

Fig. 11.1 Crowd-sourcing mode of "rrkd.cn"

delivery service, by means of integrating the idle delivery resources of the community. Under this "crowd-sourcing" mode, demand information of the last-mile delivery service of the e-commerce express is published on the Internet platform, and some franchised part-time delivery personnel would respond to specific delivery need according to their own trip schedule and free time, and provide the logistics delivery and payment services according to corresponding service process. Typical enterprises of this mode include "rrkd.cn" in the crowd-sourcing field, and the "imdada.cn" in the local living areas. The crowd-sourcing mode of "rrkd.cn" is as shown in Fig.11.1.

"rrkd.cn" is a representative company in the last-mile delivery field of China's e-commerce logistics. It signs contract with numerous "free couriers" through the mobile Internet platform of mobile APPs to integrate the human resources for the last-mile delivery. These "free couriers" include temporary workers, volunteers, interns, various part-time couriers, etc. The platform provides the required training to those "free couriers." By releasing the demand information of express delivery and drawing response from the couriers, it could accurately match the logistics demand and the supply service. At present, such service mode is still in the initial development stage, relevant laws and government regulations need to be further perfected.

11.2.2 Innovative Service Mode in Platforms

11.2.2.1 Physical Platform Mode Focused on Selling and Shopping

The selling/shopping logistics delivery platform is both a delivery platform of the last-mile delivery of e-commerce logistics, and also an online retailing O2O physical chain store, which is a retail system that links the sales function from offline to online. As a community logistics node serving B2C and C2C shopping, this type of platform has the last-mile delivery functions of express service, such as collection

folk financing, election campaign, fundraising for startups, artistic creation, free software, design and invention, scientific research and public financing. With the development of the Internet platform, the idea of "participation by all" from Crowd-funding has gradually extended to many non-financial fields, and become known as "Crowd-sourcing."

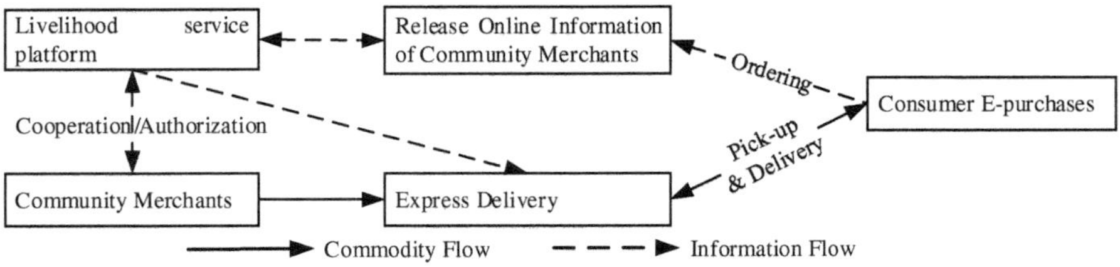

Fig. 11.2 Service mode of network platform focused on livelihood services

of goods, payment collection on delivery, return and exchange, choice options upon delivery, and community delivery. As the store of the O2O shopping, such platform has the functions of product exhibition, user experience, retail ordering, and it can also add community service business such as ATM self-service banking, tickets booking, payment of bill, and laundry operations.

At present, the more typical selling and shopping logistics delivery platforms include SF Express's "Heike", Ali's "Mail World." For example, on May 18, 2014, SF Express's "Heike" began operations nationwide in 518 so-named community's stores for online shopping services. Main services provided by these Heike stores include commodity pre-purchasing, product experience and ordering, self-served express sending and receiving, and some community services such as ATM banking, dry-cleaning, booking of airlines tickets, recharging of telephone cards and paying utilities bills. Thus the "Heike" e-commerce platform has emerged and, together with the physical chain store system, has formed a retailing business mode with linked online and offline functions. Meanwhile, this business mode has the capabilities of the last-mile delivery service of express logistics in receiving for clients, self-serve package pick-up and parcel delivery.

11.2.2.2 Network Platform Mode Focused on Livelihood Services

Generally, the platform which mainly focuses on livelihood services does not provide express package receiving service, nor have physical stores. This mode of platform is asset-light and focuses on community livelihood services in a certain area. It integrates the information of the surrounding stores and online merchants for community consumers, builds sales network platforms, and helps the consumers within the service areas who place an order to purchase products and make home delivery. Such community service mode is sometimes vividly described as the "take-out" service or the "errand" service. Figure 11.2 shows the flow of this service mode.

For example, catering takeout "ele.me" integrates the offline catering brands and online network resources on its platform to enable the consumers search for the neighboring restaurants on their mobile phones and computers, order food online and receive delivery at home. By September 2014, the "ele.me" platform had covered almost 200 cities in online food ordering service, with an average daily order

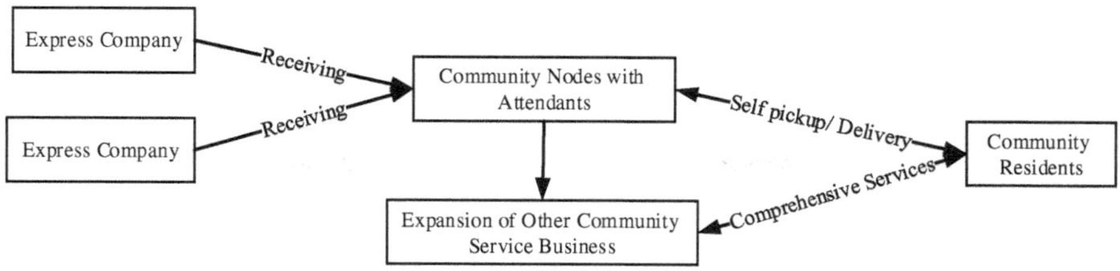

Fig. 11.3 Attended service mode of parcel receiving

of more than 1 million.[5] For another example, the last-mile delivery service of "shequ001.com" in Beijing includes online supermarkets, community purchasing by agents and community delivery, providing a variety of purchasing and express services for community life circles within a 5 km radius. Other similar platforms are SHBJ.com and the like. In addition, some large e-commerce platforms also run similar community services, such as waimai.meituan.com, waimai.baidu.com and dd.taobao.com.

11.2.2.3 The Physical Platform Focused on Parcel Receiving

Package delivery is one of the major businesses in the last-mile delivery of e-commerce logistics. Presently, express enterprises often encounter problems during the door to door service for e-commerce packages, where no addressee to receive the delivery and no reply to contacting attempt. This situation results in considerable waste of time and effort of the couriers during the delivery process. With deepening specialization of the industry, parcel receiving has gradually become a unique service content of the last-mile delivery of e-commerce logistics. By far, the parcel receipt service consists of attended mode and unattended mode.

a) Attended franchise mode

Many e-commerce enterprises, logistics express enterprises and even retailing enterprises with franchise network are plotting parcel receiving nodes for the last-mile delivery of e-commerce logistics. Generally, the attended mode for parcel receiving requires physical stores; and currently, the form of franchising various community stores (chain drug stores, laundry stores, property management and convenience stores) is commonly adopted. These receiving nodes can offer services such as receiving of package, self pickup, goods inspection, merchandise display, catalogue sales, payment collection on delivery, and return and exchange, as shown in Fig. 11.3. Enterprises engaging in the franchised mode of parcel receiving include the "Happiness Station" of Wanda Property Management, the "Tmall Service Station (the Cainiao Station)" of Tmall e-commerce, and the Family Mart supermarket chain.

[5]Taoran. Interview of Zhang Xuhao of ele.me [EB/OL]. Notes of personages in Tencent. http://tech.qq.com/zt2014/people/05.htm. 20 Oct 2014.

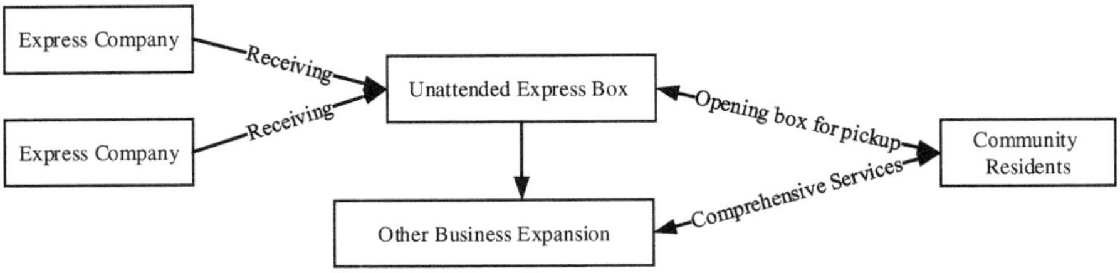

Fig. 11.4 Comprehensive service mode of the "Express Box"

For example, shouhuobao.com,[6] established in April 2011, adopts the operations mode of parcel receiving by a third party. It has developed rapidly in the parcel receiving business, and established more than 15,000 franchised parcel receiving nodes in Beijing, Shanghai, Guangzhou, and Shenzhen. These nodes can receive consumers' parcels for free and keep them in storage for five days. Generally, the busiest time of the nodes is between 6 and 8 p.m., since the consumers prefer to pick up their parcels from the location nearest to their homes along the route between workplace and home. By integrating e-commerce, express delivery and the franchised network community chain stores, "shouhuobao.com" has thus greatly improved the service efficiency and the professional level in the last-mile delivery of the ecommerce logistics.

b) Unattended "Express Box" Mode

The parcel receiving business can also take the form of "Express Box" for packages like the household mailbox. Such mode is free of attendants or physical buildings, but the setup of express box for depositing packages, designed with various sizes of pigeon holes. Compared with the attended receiving nodes, the express box is more economical but is also relatively difficult to imbed the added-value services.

The express box can have two types; one is the regular express cabinet or mail cabinet, and the other one is the so-called smart express cabinet. The smart express cabinet has the function of opening the locked box by scanning a code. The courier can send the password of the locked box to the client after making delivery to the box, and the client can use the password to open the box and retrieve the goods. The smart express cabinet service can accumulate user data, and generate revenue by charging for delivery, overdue, dispatching, and advertisement. The "Express Box" service mode is shown as in Fig.11.4.

Currently enterprises conducting express box business include spotster.net, 6box.com, and parcelcube.cn, wherein sposter.net piloted first in Chengdu and Chongqing. By October 2014, sposter.net had built up more than 7,000 nodes in the

[6] "shouhuobao.com" and its consumers consented that packages stored in the receiving location over 30 days are considered disregarded delivery by the consumers and will be returned to the senders. The shouhuobao.com service will remind the consumers of the duration of storage days. This part of the information is excerpted from the official homepage of shouhuobao.com, http://www.shouhuobao.com/. 20 May 2015.

nation's first-tier cities, high-end residential and popular online shopping areas. The express box mode can considerably improve the courier's delivery efficiency, and generate revenue through charging the express enterprises. Meanwhile, the enterprises operating the express boxes are also exploring many other profit-making modes.

11.2.3 Professional Service Mode for Specific Products

Some service platforms of the last-mile delivery focus their service contents on the professional logistics of a certain family of products, and structure the complete distribution and delivery solution according to the products' characteristics and specific demand. Currently, the products needing professional logistics service of the last-mile delivery are mainly bulky cargos, fresh and frozen foods.

11.2.3.1 Large Electrical Appliances

The last-mile delivery logistics service of large electrical appliances generally needs extended installation function; the operations process involving transport vehicles, equipment and loading and unloading also call for professional operations. For example, large appliance stores, such as GOME logistics and SUNING logistics, are expanding their logistics service capability for electrical appliances, and meanwhile emphasizing the logistics service of terminal distribution and delivery. As another example, the last-mile delivery service of "RRS Logistics" of Haier, on the one hand, coordinates the upstream supply chain and expands the online retailing, and product customization functions of the electrical appliances; on the other hand, it expands the stores, manages the SKUs according to the local demand, thus provides the last-mile delivery service mode incorporating home delivery and installation service. "RRS Logistics" acquired the strategic investment of Alibaba in December 2013, and its service has since covered the vast areas of urban and rural markets. By the end of 2014, the scale of its chain stores and franchised stores had reached over 7,600 county-level specialty stores, 26,000 rural specialty stores and 190,000 village-level contact posts.[7]

11.2.3.2 Fresh and Low-Temperature Foods

The last-mile delivery logistics service of fresh and low-temperature products requires the proper insulation and monitoring mechanisms, and meanwhile achieves timeliness of the express operations, and the promptness of delivery. In recent years,

[7] Zhao Guang. China's business mode of the "Last-mile Delivery" in 2014 [EB/OL]. Pintu net. http://www.pintu360.com/52673.html. 30 Dec 2014.

given the rapid growth of e-commerce industry for fresh foods in China, the delivery and support of the last-mile delivery played a key role, providing valuable and successful experience for the e-commerce industry for fresh foods. For example, the O2O enterprise "beequick.cn," which provides 1-h delivery of yoghurt, coffee, ice cream, pickled tidbits, petite seafood, snacks, and processed cold foods. "Beequick.cn" was online in May 2014, and grew to more than 2,000 delivery posts in Beijing, is capable of making fast fresh food delivery in the city's core areas in 30 to 120 min. In addition, this platform does not have its own full-time delivery team, but delivers by outsourcing to the "idle" resources, namely, crowd-sourcing. Another platform, "justeasy.com," mainly focuses on the service supply chain of household fresh foods in Central China; it solves the endpoint delivery and temporary depositing problems with the "electronic food box" installed in the community.

11.2.4 Professional Service Mode in Specific Regions

11.2.4.1 Delivery in Township and Rural Areas

Due to the relative low consumption capability in township and rural areas, and the constraint of the service ability of the e-commerce logistics, logistics delivery in township and rural areas becomes one of the weaker links in China's e-commerce logistics development. The township and rural areas are geographically wide with relatively low residential density, thus the logistics service of last-mile delivery needs to broaden the product categories of the platform through the establishment of physical distribution network, and meanwhile enlarge related added services, such as tailoring platforms to feature rural products and two-way logistics.

The "SSS Logistics" of Shandong Province cooperates with the "isubuy.com" platform to expand its service network, on community basis, to the township and rural areas. The main products being delivered include general merchandise, FMCG, agricultural capital goods, and small household appliances. In 2004, 106 physical logistics stores had been established in 88 townships of Taian, Shandong province. These stores provide convenience services for rural communities, such as on-site online purchasing, recharging mobile phone cards, sales of lottery tickets, paying utility bills, booking airlines and train tickets, money transfer from bank accounts, paying credit cards, and withdrawing from the rural pension, thus forming a service system combining business and logistics and adapting to the development of rural lives.

11.2.4.2 Campus Delivery

Campus and surrounding areas where college students congregate are the relatively concentrated areas of e-commerce logistics service. The last-mile delivery to the campus features concentrated delivery time (ie, students have relatively fixed school timetable), more restricted entrance regulations (ie, most schools limit the number of express vehicles entering campus). So by centralized distribution and delivery, the terminal delivery efficiency can be improved.

"imxiaomai.com" is a logistics company which mainly provides the last-mile delivery service to the campuses. It plans to make the terminal delivery of e-commerce logistics for students in 300 universities in China, in the form of on-campus business offices of physical logistics. It integrates the mainstream express companies to offer the round-the-clock self-pickup service and post-sales return and exchange services for the college students. In addition, the clients can also choose the "Mac Delivery" service to redistribute the packages to the designated on-campus locations, or utilize the intra-campus point-to-point delivery/moving service. Students who are familiar with their university can work as part-time couriers for imxiaomai.com. As a whole, imxiaomai.com provides the service mode of self-pickup, return, redistribution, and point-to-point delivery of the last-mile delivery within a specific locale, thus realizing the service integration of the last-mile delivery logistics service conforming to the characteristics of small regions.

11.3 Development Status of the Last-Mile Delivery Service of E-commerce Logistics and Its Existing Problems

Although the last-mile delivery service of China's e-commerce logistics has existed for some time, yet the diversification of its service mode emerged only in the recent 3–5 years. Many emerging O2O multiple service modes of the last-mile delivery were born just in 2014; the industry is still in a preliminary development phase, so many service modes are just being tested in some regions and not yet mature. Nevertheless, the last-mile delivery service of the e-commerce logistics has become the focus of various relevant enterprises, and new innovative programs have constantly obtained vast support from venture investment.

11.3.1 Scale of the Industry Constantly Expanded, and Innovation of Service Mode Intensified

The scale of demand for last-mile delivery service of the e-commerce logistics in China has constantly expanded. According to the State Post Bureau's statistics, the total business volume of China's express industry reached 13.96 billion in 2014, with a year-on-year growth rate of 51.9 %. The average annual growth in business volume of the nation's express industry amounted to 44.82 % from 2008 to 2014. These seven years is the very period for the rapid growth of e-commerce; a very large proportion of express packages come from the logistics demand of e-commerce, which also requires the terminal distribution and delivery services of the last-mile delivery. Meanwhile, the service mode of the last-mile delivery of China's e-commerce logistics has constantly been innovated; chain websites from express enterprises, e-commerce enterprises and community convenience stores are involved in the last-mile delivery of e-commerce logistics. The terminal distribution and delivery service of e-commerce logistics have become one business genre in industrial technological innovation and concerted business mode innovation.

11.3.2 Service Quality Yet to Be Improved While Service Complaints Remain High

The quality of the last-mile delivery service of the e-commerce express enterprises in China needs much improvement. According to the State Post Bureau's statistics, in 2014, the accepted valid complaints for the express services involving delivery service, posting service, and collection on delivery, which are directly related to the last-mile delivery are 822,000, 8,000 and 2,100 cases, respectively. The numbers of complaints of these problem issues increased as compared with the relevant valid complaints in 2013; the growth rates of the first two items are relatively higher, reaching 43.2 % and 37.4 %, respectively. Compared with those in 2013, the complaint rates of the delivery business and posting business grew faster than the overall complaint issues. Table 11.2 shows more details.

There are three causes for the relatively low quality and the relatively high complaint rate of the last-mile delivery service of express logistics. First, the current avenue of competition in the express industry is focused mainly on price, so the firm's profit margin is severely compressed. The couriers are paid by pieces delivered, so their primary objective is to finish the assigned deliveries. Second, in the process of logistics distribution and delivery, diversity of consumers' demands and nonstandard services create difficulty in coordination and operation. For example, under general service mode, the courier would have to contact the consumer for every single package; the cost of coordinating a mutually agreed meeting time and location is huge. Third, the subjective and objective conditions, such as traffic congestion in large cities, inadequate delivery facilities of the last-mile delivery, unreasonable service mode, and low quality of the express personnel, can result in high rate of service complaints.

11.3.3 Regional Development Extremely Imbalanced Though Service Modes in First-Tier Cities Relatively Abundant

The development level of the last-mile delivery service of e-commerce logistics among different cities and regions is extremely imbalanced. In some first-tier cities, such as Beijing, Shanghai, Guangzhou and Shenzhen, there are more modes of

Table 11.2 Complaints on China's express business for 2013–2014

Items (Unit)	Delivery Service	Delay	Loss/Deficit	Damage	Receiving	Illegal Charges	Collection on Delivery	Other	Total
2013 (Piece)	57,412	85,164	30,921	12,562	5800	1691	2046	450	196,046
Total Percentage (%)	29.3	43.4	15.8	6.4	3.0	0.9	1.0	0.2	100.0
2014 (Piece)	82,188	82,988	40,679	15,551	7969	1781	2075	1466	234,697
Total Percentage (%)	35.0	35.4	17.3	6.6	3.4	0.8	0.9	0.6	100.0
Year-on-year Increase (Piece)	24,776	−2176	9785	2989	2169	90	29	1016	38,651
Year-on-year Growth (%)	43.2	−2.6	31.6	23.8	37.4	5.3	1.4	225.8	19.7

Source: Compiled from statistics of the State Post Bureau of PRC (2013, 2014) http://www.spb.gov.cn/xytj/tjxx/

last-mile delivery service, the proportion of self-operated logistics among e-commerce platforms is relatively high, and the delivery nodes and investment in transport resources by express enterprises are relatively ample. In the second-tier and third-tier cities and township and rural areas in which the development level is relatively backward, the last-mile delivery service of e-commerce logistics is also relatively backward. The emphasis of network development in e-commerce express is still on network coverage. Consequently, investment in service differentiation and value-added service by on online retailing and logistics enterprises are severely deficient.

In addition, noticeable differences in service efficiency and service quality may exist among express enterprises which undertake the last-mile delivery service in the same region or even the same city, due to different operational modes and management capabilities. Demands for last-mile delivery service are dissimilar, and the development of service level is also distinctive for different types of communities, such as business districts in city centers, schools, enterprises, residential districts in old town, and new city districts. Generally, in communities with dense population and frequent online shopping activities, there will be more optional last-mile delivery service modes with higher service quality. Conversely, in communities which feature traditional retailing and low demand, their last-mile delivery service mode will likely be simple and backward.

11.3.4 Market Segments Gradually Taking Shape, and Professional Level of Service Continually Improved

The last-mile delivery service of e-commerce logistics has also been developing towards segmentation and professionalization. First of all, it has gradually expanded in various sectors involving intra-city delivery, integrated warehousing and distribution, terminal delivery and parcel delivery, and eventually fostering service providers with market segmentation and professionalism. Some large e-commerce express enterprises, such as "STO," "YTO," "ZTO," and "YUNDA," can provide complete city delivery solutions with "integrated warehousing and distribution." Some regional ground delivery enterprises, such as "ewinshine.com" and "fyps.cn," focus on the construction of network nodes for urban delivery and urban depots.

Next, the last-mile delivery service has gradually grown in professionalization regarding to various service products and different community types. Different service products, such as the "4000121777.com," and the "cnznzy.com" platforms, focus on hot pot take-out and home delivery in the community; the "edaixi.com" platform provides door-to-door laundry pickup, cleaning, and delivery service. And even some personal service industries are also exploring the O2O door-to-door service to reduce the customers' travel. Examples are the community housekeeping services of "ayibang.com," and the door-to-door nail care of "helijia.com."[8] For different community types, the "daojia.com.cn" platform mainly offers food takeout services to the high-end and mid-end communities. Some other innovative services focus on office buildings as their target of takeout logistics.

Finally, some last-mile delivery enterprises also culminate the delivery service in public places with high traffic flows. For example, the "qingniancaijun.com" platform in Beijing provides consumers with semi-finished processed foodstuff; they clean and cut the foodstuff ordered by the consumers, prepare needed seasonings for cooking the dish, then take the ready-to-cook food to the Beijing subway station to make the last-mile delivery to the customers. By choosing the terminal delivery location in the convenient passageway, it can accurately serve the specific crowds at specific time, and in specific sites.

11.3.5 Enterprises in Competition Diversified, and Profitable Mode Yet to Be Explored

With the diversification of the last-mile delivery service mode in China's e-commerce logistics, the type and number of enterprises joining the market competition are continually increasing; many enterprises enter the logistics business of the last-mile delivery by stepping out of their main business arena. These participants include e-commerce platforms like "Jingdong.com" and "Tmall.com," community property management enterprises like the "Happiness Station," and also convenience stores in the community, such as various retail chain stores, drugstores and laundry stores. The last-mile delivery service markets in certain cities and some high-end communities have entered the stage of competition among numerous enterprises.

Even though the community last-mile delivery service market presents a huge potential, yet at present, most enterprises partaking in market competition are laying the groundwork of service network layout, and relying on the support from venture investment. Low profit margin of the community e-commerce logistics service, high cost in physical stores, ill-repute resulted from cumbersome non-standard services, integration capability towards the product's upstream supply chain, drawing and keeping consumers, and other problems all become factors which constrain the betterment of profit. At present, most these enterprises are still exploring the direct and indirect modes of making a profit.

[8] Strictly speaking, these O2O door-to-door services do not entail much "Last-mile Delivery" distribution and delivery of products, and mainly provide the O2O door-to-door services. But some O2O services also involve some form of delivery service in bring along or recommending relevant products.

11.3.6 Supervision System of the Government Yet to Be Perfected

The legal system of the last-mile delivery service in China's e-commerce logistics is still not well-established, and the relevant government regulations, administrative measures and standards are yet to be perfected. For example, vehicle accessing system of city logistics delivery is closely related to the last-mile delivery service, but the governing of small delivery vehicles (such as three-wheeled electric vehicles) regarding their road-right, parking and accessing are not standardized in different cities. For cost-saving considerations, enterprises use many vehicles with insufficient safety features, thus resulting in traffic safety hazard in the cities. For another example, the explicit administrative measures and legal basis for resolving problems of delivering prohibited products, insurance recompense on express articles, and disputes on collection on goods in the last-mile delivery service are not yet instituted.

11.4 Development Trend in the Last-Mile Delivery Service of E-commerce Logistics

The last-mile delivery service of China's e-commerce logistics has developed rapidly in recent years; new technologies, new equipment and new business modes are continually emerging. Under the backdrop of accelerated promotion of new technological revolution in areas of Internet, mobile e-commerce, Big Data and intellectualization, the developmental level of China's e-commerce logistics will be elevated accordingly, and the last-mile delivery logistics service will also grow towards maturity and perfection in time.

11.4.1 Cross-Sector Cooperation of Enterprises and Further Acceleration of Service Integration

With the diversification of the last-mile delivery service modes, cross-sector cooperation and service integration among enterprises possessing different service resources/capabilities will be further accelerated. The goal of customers choosing varied service modes according to their specific requirements will gradually come to pass. When the industry's market is developed to a certain degree, regional market integration will be inevitable. Network resources and the service level of large enterprises will be enhanced, and the monopoly power of the market will also be expanded thereupon. Due to differences of cities, communities, people and products inherent in the last-mile delivery service, large enterprises will also be compelled to elevate their own diversified service capability.

11.4.2 Professionalization and Localization Becoming Important for Enterprises

The last-mile delivery service of e-commerce logistics is professionally inclined, so service barrier and core competitiveness of the leading enterprises are relatively easy to establish. Therefore, enterprises having professional last-mile delivery service will be able to preserve reasonably stable breathing space. Enterprises serving specific cities and even specific communities are likely to attain core competitiveness through long-term accrued customer relations and experience, and become an essential component of the city community service systems. For example, local e-commerce and logistics service platform, ie "taocz.com," focuses their targeted market on several communities in Changzhou City. By strengthening the close relationship with the community residents, the social attributes of last-mile delivery service are increased, so that the enterprise has built up a favorable reputation among the locals, thereby, forging a unique competitiveness of local services.

In the future, the crucial components of enterprise competitiveness in the last-mile delivery service competition of e-commerce logistics are to serve the exceedingly personalized consumer demand, form solid consumer loyalty and community reputation, rely on repeat purchase of local customers, and savings on the communication costs.

11.4.3 Unmanned Aerial Vehicle (UAV) and Other New Technologies May Subvert Service Modes

The development of emerging technologies will be a deciding factor for the selection of business modes by the enterprises. The future development and application of new technologies, such as UAV, 3D printing, industry 4.0, Big Data and cloud computing, may subvert today's last-mile delivery service mode of e-commerce logistics. For example, Amazon has begun to explore the terminal distribution and delivery services of "Distribution Vehicles+UAV," namely, the transport vehicles take up the terminal distribution of the main roads, and then the UAV is sent to finish the delivery of the last part. For distribution cost consideration, the power consumption cost of UAV is about two cents per mile, but the fuel cost of an ordinary delivery vehicle is about 50 cents per mile. For another example, the 3D printing technology can print some simple products directly according to the consumer needs, so that the logistics demand and the last-mile delivery service for these types of products could possibly be reduced by a large margin.

11.5 Increasingly Prominent Requirement of Energy Conservation and Environmental Protection

Low carbon emission and environmental protection are one of the development directions of global industry, and cannot be overlooked by the last-mile delivery service industry of e-commerce logistics. Many operations of the last-mile delivery are carried out in cities; as a result, the impact of pollutant discharge from vehicles and energy consumption are multiplied. At present, many developed countries are undertaking the "Green Express" action, attempting and exploring green packages, new energy vehicles, energy conservation construction, and participation in carbon compensation plans. For example, the U.S. Postal Service has enforced the use of green package materials and implemented using electric vehicles for community express delivery. China has also begun the adoption and promotion of new energy vehicles in the express industry. In the future, green express is likely to be the fundamental requirement for the last-mile delivery service of logistics.

11.6 Summary

As a whole, the last-mile delivery service of China's e-commerce logistics has developed under the backdrop of e-commerce express, demand personalization, and distribution of logistics network extending towards urban community. Its characteristics include personalization, high level of customized service, imbalanced development in time and space, two-way logistics in forward and backward directions, and plentiful room for development of value-added services. The last-mile delivery service of China's e-commerce logistics can be categorized into four types, ie, innovative service mode in express enterprises, innovative service mode in platforms, professional service mode in specific products, and professional service mode in specific regions. Generally, its development trend features continuously expanding market size, increasingly diversified service modes, gradually formed market segments, and constantly emerging service innovations. Meanwhile, the last-mile delivery service of e-commerce logistics also confronts many problems, such as high service complaints, imbalanced regional development, profitable mode still to be explored, and government's administrative measures yet to be improved. Regarding the future development trend, the last-mile delivery service of China's e-commerce logistics will be developing towards the areas of cross-sector cooperation, service integration, professionalization, localization, adoption of new technologies, and conformity to the requirements of energy conservation and environmental protection.

References

Taoran (2014) Interview of Zhang Xuhao of ele.me. Notes of personages in Tencent. http://tech.qq.com/zt2014/people/05.htm. Accessed 20 Oct 2014.

Zhao Guang (2014) China's business mode of the "Last-mile Delivery" in 2014. Pintu net. http://www.pintu360.com/52673.html. Accessed 30 Dec 2014

Chapter 12
Development of Cold Chain Logistics in China

Xiang Li

With the economic development of the society and the improvement of residents' consumption level, the fresh product market in China has grown yearly and product quality assurance has also become the concern of the whole society. As a core entity of fresh product circulation, cold chain logistics has been developed rapidly. This is of great importance in safeguarding the consumption safety of food and drug, improving people's quality of life, and promoting the transformation and upgrading of the logistics industry in China.

This chapter presents the development of cold chain logistics in China in four sections. Section 12.1 introduces the operating subjects, the linkages, processes and features of cold chain logistics. Section 12.2 reports the development environment of cold chain logistics in China from three aspects: market, policy, and technical environment. Section 12.3 describes the development status of cold chain logistics in China, including logistics demand, infrastructure, rail and air transportation, third-party cold chain enterprises, leading enterprises, and e-business cold chain. Section 12.4 discusses the existing problems and development trend of cold chain logistics in China.

12.1 Overview of Cold Chain Logistics

Cold chain logistics is a professional logistics mold to always keep chilled and frozen food, drug and other goods under prescribed low temperature during the production, storage, transportation, and sales stages prior to delivering them to the consumers, in order to ensure their quality and reduce spoilage. Cold chain logistics

X. Li (✉)
Logistics Research Center, Nankai University, Tianjin, People's Republic of China
e-mail: xiangli@nankai.edu.cn

L. Wang et al. (eds.), *Contemporary Logistics in China*, Current Chinese Economic Report Series, DOI 10.1007/978-981-10-1052-1_12

applies to agricultural products (fruits and vegetables, meat, poultry, eggs, aquatic products, dairy products, flowers, etc.), processed foods (quick-frozen food, packaged cooked food and raw materials for fast food, etc.), and other special goods (temperature-controlled pharmaceuticals and some hazardous chemicals).

12.1.1 Operating Subjects, Links and Processes of Cold China Logistics

12.1.1.1 Operating Subjects

Different cold chain products such as primary agricultural products and processed foods have different circulation modes, which involve different operating subjects. Generally speaking, operating subjects of cold chain logistics include producing and processing enterprises, wholesale enterprises, retail enterprises, platform e-commerce enterprises and third-party cold chain logistics enterprises.

12.1.1.2 Operating Links

Major operating links of cold chain logistics include pre-cooling, freezing/processing, cold chain storage, cold chain transportation and distribution, and cold chain sale.

First, pre-cooling is the starting point of the cold chain. Most agricultural products, such as fruits, vegetables and meat, will be pre-cooled before being sent to the freezing/processing link and storage link. Pre-cooling facilities and devices mainly include various pre-cooling machines and pre-cooling and freshness-preserving storehouses.

Second, freezing/processing is a link to improve the usage effectiveness of cold chain products through a series of processing works under low temperature, and mainly involves various fast-freezing, cooling and low-temperature processing equipments.

Third, cold chain storage is a link to store the products under specified low temperature with refrigerated warehouse and relevant equipment therein. Different products have deferent requirements for storage temperature; ice cream, for example, and other frozen food must be kept at a refrigerated storage temperature of −18 °C or below, and fruits and vegetables must be kept at a temperature of 0 °C–8 °C (Ye Haiyan 2007).

Fourth, cold chain transportation and distribution plays a key role in always keeping the products under low temperature during the process of physical displacement. Long-distance transportation includes highway, rail and air transportation, while short-distance transportation tends to rely on highways. During the transportation process, it is necessary to utilize high-performance cold chain transportation tools, strictly follow temperature control standards and preferably use a same

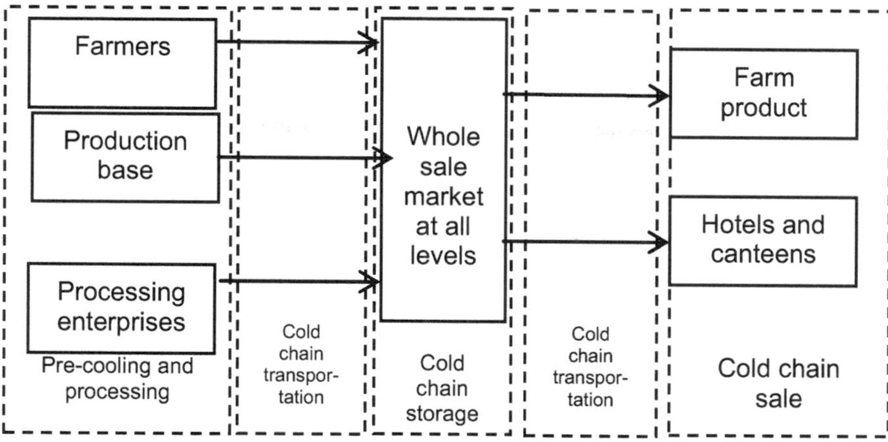

Fig. 12.1 Process of cold chain logistics in wholesale market-based circulation channel

vehicle or multi-modal transport with refrigerated containers for door-to-door transportation.

Finally, the link of cold chain sale includes cold storage, exhibition and sale of products in wholesale and retail link, generally executed by wholesalers or retailers. In this link, a lot of refrigerated display cabinets and storage houses will be used.

12.1.1.3 Operating Processes

Currently, the main circulation channels of agricultural products, food and other cold chain food include wholesale market, supermarket and e-commerce platform in China, and the cold chain logistics has different processes in different circulation channels.

(1) In the wholesale market-based circulation channel, the products may be first pre-cooled and cold chain processed by farmers, bases or producing and processing enterprises, then transported to wholesale markets at all levels through cold chain transportation and refrigerated within cold storage facilities in wholesale markets and finally distributed to farm product fairs or catering enterprises to meet consumers' demands, as shown in Fig.12.1. Due to the large number of logistics links in this channel and still imperfect cold chain hardware facilities in most wholesale markets in China, cold chain logistics' efficiency is relatively low and products deterioration is relatively high in the process of logistics.
(2) In the supermarket circulation channel, the products, after pre-cooling and processing, will be sent to the supermarket terminals directly or through distribution centers of cold chain logistics, as shown in Fig. 12.2. This process reduces some logistics links and product deterioration along the cold chain, better satisfying the logistics demand of "contract farming," "connecting farmers' coop-

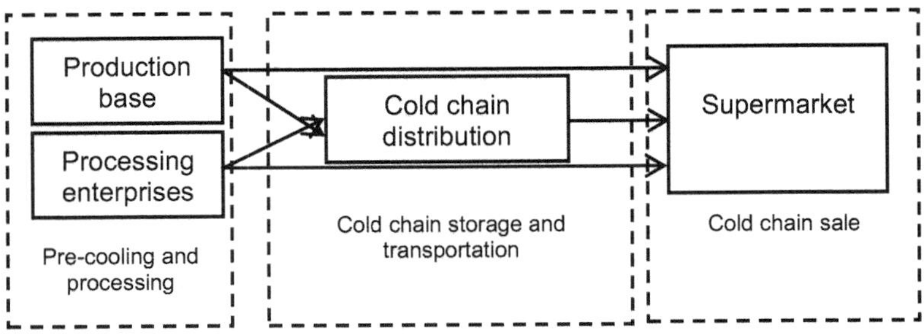

Fig. 12.2 Process of cold chain logistics in supermarket circulation channel

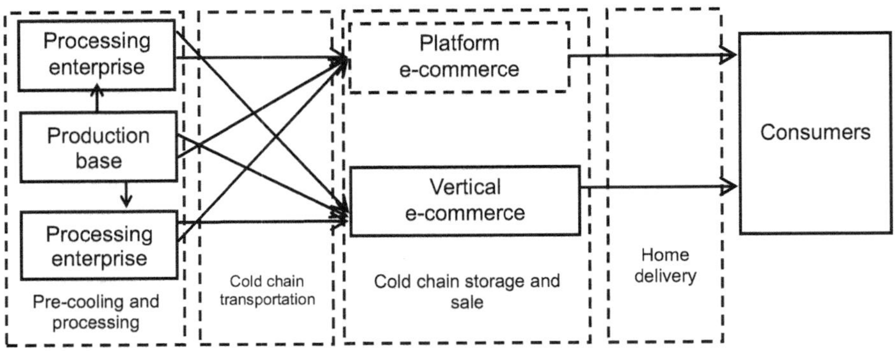

Fig. 12.3 Process of Cold Chain Logistics in E-commerce Circulation Channel

eratives and supermarkets" and other new circulation modes of agricultural products.

(3) In the e-commerce channel, the products will be sold to consumers through e-commerce websites. On the one hand, platform e-commerce is only an online marketing and trading platform; fresh products will be directly sent and delivered to consumers from producing origins by suppliers; the cold chain logistics is often delegated to the third-party logistics providers. On the other hand, vertical e-commerce related to self-operated fresh product businesses often participate in part of the operations of cold chain logistics through direct purchase from producing origins or self-built bases, as shown in Fig. 12.3. In addition, cold chain home delivery is added to circulation mode of e-commerce channel as a terminal logistics link, which is different from the former two channels in which consumers go to farm product fairs or supermarkets to buy products by themselves.

12.1.2 Characteristics of Cold Chain Logistics

12.1.2.1 Influence on Product Quality Safety and National Economy and the People's Livelihood

To prevent/reduce deterioration of fresh agricultural products, foods, drugs and other products, it is required that they be always kept under specified low temperature in order to realize a cold chain logistics system which enclose the entire circulation process. Any broken/out of control link may result in quality safety problem of cold chain products or even give rise to serious public safety incident. Therefore, cold chain logistics is of great importance in maintaining consumption safety, national economy and the people's livelihood.

12.1.2.2 Huge Investment in System Construction and High Operating Cost

Construction cost of cold chain infrastructure is often many times that of common logistics facilities, so huge investment is needed. Moreover, power consumption is constantly needed in the operation of cold chain logistics in order to safeguard the low temperature the products need, so the energy consumption cost is relatively high. Table 12.1 and Table 12.2 respectively show the cost of self-operated and

Table 12.1 Cost of self-operated cold chain logistics

Cost type	Cost contents		
Storage cost	Construction cost	Construction cost of refrigerated warehouses from 50,000 M^2 to 200,000 M^2 is 45 million RMB ~ 180 million RMB higher than that of ordinary warehouse.	
	Rent	Take Shanghai as an example, cold storage charge is 2.82 RMB/(ton-day).	
	Power consumption, 80 % consumed by refrigerating system (unit: Kw · h/ton-day)	Cold storage of frozen product	Medium and large refrigerated warehouse: 0.4–0.5
			Small refrigerated warehouse: 0.6–0.9
		Cold storage of chilled product	Medium and large refrigerated warehouse: 0.9
			Small refrigerated warehouse: 1.4–1.5
Transportation cost	Oil consumption	Fuel consumption per refrigerator truck will generally increase by 30 %–40 %.	
	Personnel salary	Chilled product: 20 RMB/(piece · person · day); ordinary product: (regardless of the number of products) 60–70 RMB/(person · day).	

Source: Compiled from the *Report of the Progress of China Fresh Product E-commerce*, published by the China Merchants Securities. http://www.logclub.com/thread-150284-1-1.html

Table 12.2 Comparison between cost of outsourced cold chain logistics and cost of outsourced logistics under normal temperature

Supply link	Classification	Example	Cost comparison
Trunk transportation	Normal temperature	From Shanghai to Beijing: 0.5–0.6 RMB/kg	Trunk transportation cost of chilled product is 4–5 times that of products under normal temperature.
		From Shanghai to Qingdao: 0.35–0.4 RMB/kg	
	Cold storage	From Shanghai to Beijing: 1.9 RMB/kg	
		From Shanghai to Qingdao: 2.1 RMB/kg	
Storage	Normal temperature	First-tier cities: 30 RMB/mon · m^2	Storage cost of chilled product is 4 times that of products under normal temperature.
		Second-tier cities: Max. 24 RMB/(mon · m^2)	
	Cold storage	First-tier cities: Max. 120 RMB/(mon · m^2)	
		Second-tier cities: Max. 100 RMB/(mon · m^2)	
Home delivery	SF Express	Calculation of chilled product freight is the same as that of products under normal temperature, with additional 5 RMB/order.	Home delivery cost of chilled product is charged a certain amount over that of products under normal temperature.
	YAMATO	Calculation of chilled product freight is the same as that of products under normal temperature, with additional 10 RMB/order, 20 RMB/order or 30 RMB/order according to package size.	

Source: Compiled from the *Report of the Progress of China Fresh Product E-commerce,* published by the China Merchants Securities. http://www.logclub.com/thread-150284-1-1.html

outsourced cold chain logistics, and the comparison of logistics costs under chilled and ordinary temperatures.

12.1.2.3 Higher Proficiency Requirements for Cold Chain Logistics Operations

In addition to professional cold chain facilities and devices, effective operation of cold chain logistics also proposes a relatively higher requirement for techniques, management and personnel support. For example, to maintain strict temperature control in each link so as to avoid chain breakage, the cold chain system needs precise temperature control, monitoring technology and alarm sub-system to realize tracking and tracing. In order to realize collaborative and efficient operation of the entire cold chain logistics system, great efforts must be given on coordination, organization and lean management of each logistics operating unit. In addition, management personnel of cold chain logistics not only needs to have good operational capability to run the logistics and supply chain, but also possesses a certain level of knowledge in agriculture science and food science.

12.2 Development Environment of Cold Chain Logistics in China

A sound development environment is the basis to achieve rapid development of cold chain logistics. On the one hand, cold chain logistics has received more attention by governments at all levels in China, and relevant polices and plans have been released in succession. On the other hand, Chinese citizens have raised increasingly higher requirement for consumption level and cold chain logistics, resulting in constantly improved market environment for cold chain products. In addition, research and application of the technologies related to all cold chain links have advanced continually, providing ample space for the development of cold chain logistics.

12.2.1 Market Environment of Cold Chain Logistics Development

12.2.1.1 Expanded Production Scale of Cold Chain Products

Fresh agricultural products and foods have a massive output volume in China and enjoy a constantly growing market demand, creating a vibrant demand space for cold chain logistics. In 2013, output of fruits and vegetables were respectively 250 million tons and 730 million tons in China, with year-on-year growth of 4.3 % and 5.8 %; output of aquatic products was 61.72 million tons, with year-on-year growth of 4.5 %; output of meat was 85.36 million tons, with year-on-year growth of 1.8 %; output of eggs was 28.76 million tons, with year-on-year growth of 0.5 %; output of quick-frozen rice and flour products was 5.55 million tons, with year-on-year growth of 18.2 % (National Bureau of Statistics of China 2014). Meanwhile, market scale of the main object of medicine cold chain logistics – temperature-controlled pharmaceuticals is also expanding rapidly. For example, in 2013, the revenue of blood products and vaccines were respectively 16.2 billion RMB and 13.7 billion RMB, up by 28.6 % and 30.5 % from that of the previous year (Cold Chain Logistics Committee of China Federation of Logistics and Purchase 2014a).

Table 12.3 Chinese Urban residents' per capita consumption of fresh products

Item (kg)	2000	2005	2010	2011	2012
Pork	16.7	20.2	20.7	20.6	21.2
Poultry	5.4	9.0	10.2	10.6	10.8
Fresh egg	10.2	10.4	10.0	10.1	10.5
Aquatic product	11.7	12.6	15.2	14.6	15.2
Fresh milk	9.9	17.9	14.0	13.7	14.0

Source: Compiled from the *China Statistical Yearbook* (2013, 2014), published by the National Bureau of Statistics of China

12.2.1.2 Elevated Citizens' Consumption Level

Chinese people's materials demand has been gradually transformed from the subsistence type to the nutritional management type, which has driven up the per capita consumption level of fresh products. In addition, the consumption proportion of chilled meat[1] mainly sold in supermarket is continually enlarged; the consumption scale of pasteurized milk and yogurt, which have relatively high cold chain requirement, has risen year by year, and the amount of certain imported foreign fresh agricultural products are also constantly increased. Chinese urban residents' per capita consumption of fresh product is as shown in Table 12.3.

12.2.1.3 Rapidly Developed Fresh Products E-commerce

With modern IT development and the change in residents' consumption pattern, e-commerce of fresh products has shown a rapid development momentum in China. Comprehensive online malls such as Tmall.com, Jd.com, Yhd.com, Suning.com and vertical food e-commerce enterprises such as Womai.com, Tootoo.cn and Benlai.com have fully charged into the fresh products market, and the Internet trading volume of fresh products has been bloated significantly. Take Alibaba.com[2] as

[1] Chilled meat means fresh meat which comes from slaughtered livestock, being quickly cooled down to 0-4 °C within 24 h, is always kept between 0-4 °C in all subsequent processing, circulation and sale process. Chilled meat features high sanitary quality, great taste, easy processing and cooking. However, compared with frozen meat (which is thawed for sale after cold storage under −18 °C), chilled meat has a much stricter process requirement, higher price and a shorter shelf life.

[2] Data of Alibaba Group, which is the largest e-commerce platform enterprise in China, have obvious significant representativeness. The Alibaba platform consistsof a C2C platform (Taobao.com), a B2C platform (Tmall.com), and a B2B platform for Internet wholesale trade (1688.com). In 2013, sales volume of agricultural products of the former two platforms accounted for 97.25 % of the total, and third platform accounted for 2.71 % of the total.

an example, in 2013, sales amount of agricultural products was 41.95 billion RMB, up by 112.2 % from the previous year, of which fresh agricultural products (including aquatic products, meat, fresh vegetables, and cooked foods) enjoy the highest growth rate, with a year-on-year growth of 194.6 % (Alibaba Group Research Center 2014). The momentous growth provides a huge development space for cold chain logistics, especially in cold chain home delivery.

12.2.2 *Policy Environment of Cold Chain Logistics Development*

12.2.2.1 Issuance of Policies and Plans Related to Cold Chain Logistics

The Chinese government has paid considerable attention to the development of cold chain logistics. In June 2010, the National Development and Reform Commission (NDRC) issued the *Planning for Cold Chain Logistics Development*, which is the first plan specifically instituted for cold chain logistics in China. In September 2014, the State Council released the *Medium and Long-term Plan for Logistics Industry (2014–2020)*, which called for accelerating the research and development of professional logistics equipment for food cold chain, and strengthening the construction of cold chain logistics facilities related to fresh agricultural products. In December 2014, the NDRC, together with nine ministries including the Ministry of Finance and the Ministry of Commerce, released the *Guidance on further Accelerating Healthy Development of Cold Chain Transportation and Logistics Enterprises*, putting forward nine suggestions and measures for accelerating the development of cold chain transportation and logistics enterprises: (1) to strongly improve the scaled and intensive development level of cold chain transportation; (2) to strengthen the construction of cold chain logistics infrastructures; (3) to perfect the standards system of cold chain transportation and logistics; (4) to actively promote the informatization construction of cold chain transportation and logistics; (5) to strongly develop joint distribution and other advanced distribution modes; (6) to optimize traffic management measures of urban distribution vehicles; (7) to enforce and improve industry supervision; (8) to enhance finance and taxation policy support; (9) to empower the functions of the industry associations.

In addition, the Ministry of Commerce and the Taiwan Affairs Office of the State Council designated Tianjin and Xiamen as the first batch of pilot cities for cross-strait cooperation in cold chain logistics in 2011. By 2014, both sides of the Straits had signed 34 letters of intent for cooperation. In August 2014, Kunshan, Beijing and Wuhan were listed as the second batch of pilot cities for cross-strait cooperation in cold chain logistics industry.

12.2.2.2 Accelerated Standardization of Cold Chain Logistics

China has made a notable progress in standardization of cold chain logistics; logistics standards at all levels have been well established and improved. The standards and specifications released after 2011 mainly include: (1) basic standards; for example, the *Classification and Basic Specification for Cold Chain Logistics (GB/T 28577–20122)*, the *Management Requirements for Traceability in Food Cold Chain Logistics (GB/T 28843–2012)*, the *Requirements and Capability Evaluation Indicators of Cold Chain Services for Logistics Enterprises (GB/T 31086–2014)*, etc.; (2) cold chain industry and equipment standards; for example, the *Specifications for Cold Storage Operation Practice (GB/T 31078–2014)*, the *Specifications for Cold Chain Transportation of Goods by Air (MH/T 1058–2014)*, the *Requirements and Test Methods of Refrigerated Vehicle for Road Transportation of Perishable Food and Biological Product (GB 29753–2013)*, etc.; (3) special industry standards for meat, aquatic products, fruits and vegetables and medicines; for example, the *Practices for Cold Chain Transportation of Livestock & Poultry Meat (GB/T 28640–2012)*, the *Service Specifications of Cold Chain Logistics for Aquatic Products (GB/T 31080–2014)*, the *Guidelines for Storage and Transportation Technique of Vegetables (GB/T 26432–2010)*, the *Operational Specifications for Medicines Cold Chain Logistics (GB/T 28842–2012)*, etc.

12.2.3 Technical Environment of Cold Chain Logistics Development

As a system engineering, cold chain logistics contains and involves extensive and complex technical fields, mainly including cold storage technology serving product logistics and the transfer processes, preservation and packaging technology for maintaining quality and safety, energy saving technology serving logistics facilities, and cold chain monitoring, tracing, and information technology for carrying out efficient logistics operations.

Over years of development, China has made considerable progress in cold chain logistics technology; it has constantly enhanced its ability in independent research and development, as well as innovation of cold storage, preservation, packaging and energy saving technology. It has begun to extend some advanced cold chain facility technologies and management technologies to an even wider scope. For example, in 2009, China Railway Tielong Container Logistics Co., Ltd. independently researched and developed the 45-ft refrigerated container for rail transportation, by integrating internationally matured container technology and advanced refrigeration technology with domestic rail and road transportation conditions. The newly designed refrigerated container was tested at the end of that year, and has now become a main tool for container cold chain transportation in China. As another example, some progress has been made in the National Science & Technology Pillar

Program during the 12th Five-year Plan Period, in the "Safe and Low-carbon Logistics Technology and Equipment for Fresh Product" program, 43 core technologies of cold chain logistics have been developed and applied to 34 agricultural product logistics bases for demonstration purposes.

12.3 Current Development Status of Cold Chain Logistics in China

Cold chain logistics has a bright market prospects in China and its development level has continually been improved. With a growing demand in cold chain logistics, China is incessantly increasing its investment in cold chain logistics infrastructures, such that the transportation system of cold chain logistics has been developed rapidly. At the same time, large food processing, circulation and retail enterprises and fresh product e-commerce enterprises have both paid great attention to the construction of cold chain logistics system, and third-party cold chain logistics enterprises have also been enhancing their service functions.

12.3.1 Increasing Demand of Cold Chain Logistics

The perpetually increasing living standard of Chinese citizens gives rise to a vibrant market demand of cold chain logistics. In 2013, total potential value of cold chain logistics (Cold Chain Logistics Committee of China Federation of Logistics and Purchase 2015) in China was 3.26 trillion RMB, with a year-on-year growth of 14.8 %; and the scale of cold chain demand was nearly 90 million tons.

By category, the total potential value of cold chain logistics of vegetables was the highest, reaching 1,456 billion RMB, accounting for 44.8 %; the total potential value of cold chain logistics of aquatic products was 971.46 billion RMB, accounting for 29.9 %; the total potential value of cold chain logistics for meat and meat products was 408.83 billion RMB, accounting for 12.6 %; that of melons and fruits and their products was 146.65 billion RMB, accounting for 4.5 %; that of milk and its products was 71.89 billion RMB, accounting for 2.2 %; that of quick-frozen foods and frozen drinks was 64.98 billion RMB and 41.47 billion RMB, accounting for 2.0 % and 1.3 % respectively (Cold Chain Logistics Committee of China Federation of Logistics and Purchase 2014a). The total potential values of cold chain logistics of different products are as shown in Fig. 12.4.

Moreover, total value of cold chain logistics of imported foods has grown sharply in China. In 2013, total value of cold chain logistics of imported food reached 91.14 billion RMB, with year-on-year growth of 21.1 %, and accounting for 2.8 % of the total value of cold chain logistics. Growth in the volume of imported cold chain

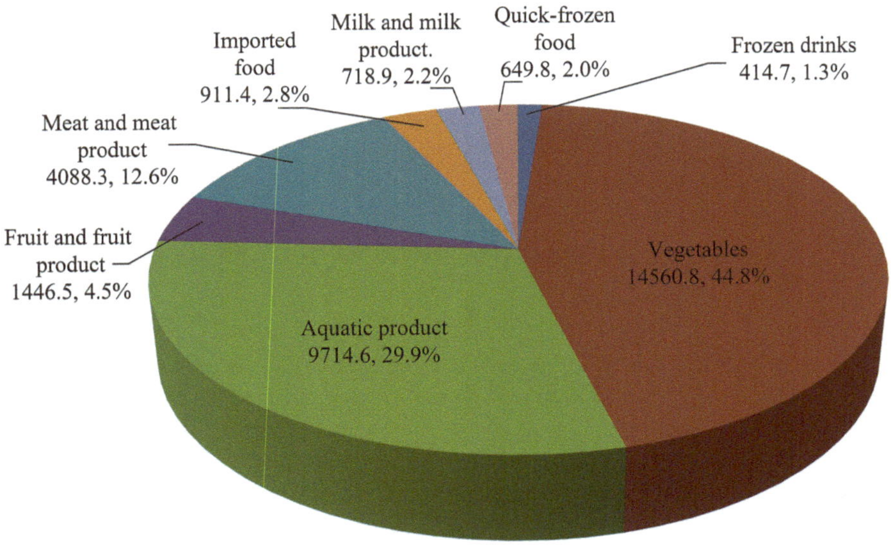

Fig. 12.4 Total potential values (Billion RMB) of cold chain logistics of different products in 2013
Source: Compiled from the Report on China's Cold Cain Logistics Development of 2014, by Cold Chain Logistics Committee of China Federation of Logistics and Purchase, China Logistics Technology Association, National Engineering Research Center for Agricultural Products Logistics, published by China Fortune Press, Beijing

food creates a favorable development space for cold chain logistics around the ports.

12.3.2 Heightened Efforts in Cold Chain Infrastructure Construction

12.3.2.1 Cold Chain Logistics Park

In recent years, China has accelerated the pace in building cold chain logistics parks. According to incomplete statistics, China constructed over 40 key cold chain logistics parks (including those completed, begun and contracted) in 2014, involving investment of over 55 billion RMB and adding more than 1.8 million tons of refrigerated warehouse capacities. Of the total investment, over 20 billion RMB was expended in North China (Beijing, Tianjin, Shandong Province, etc.), accounting for more than 1/3 of the national investment. Of the total, about 5 billion RMB was invested in Northeast China, about 6 billion RMB in South China and East

China (Guangdong Province, Shanghai, Zhejiang, etc.), about 5 billion RMB in Southwest China (Chongqing, Sichuan Province, etc.), and about 5 billion RMB in Central China (Anhui Province and Hubei Province, etc.) (Cold Chain Logistics Committee of China Federation of Logistics and Purchase 2015).

Over years of construction, the cold chain logistics park network has taken shape in China, and China has made efforts to continually promote its scale and improve its performance. For example, in Circum-Bohai Area, a cold chain logistic network bunching 14 cold chain logistics parks as core nodes has been built, radiating throughout the entire area of Beijing, Tianjin, Hebei Province and Northeast China, and rendering all-around service including cold chain storage, logistics distribution, circulation and processing, and trade exhibition.

12.3.2.2 Refrigerated Warehouse

The gross scale of refrigerated warehouses has continually been on the rise in China in recent years. In 2014, total storage capacity of all refrigerated warehouses in China was 33.2 million tons, with year-on-year growth of 36.9 %; among them, public refrigerated warehouses accounted for the most part of newly constructed refrigerated warehouses. Table 12.4 shows the mega-sized refrigerated warehouses newly built in past two years with storage capacity of over 100 thousand tons or 500 thousand cubic meters.

12.3.2.3 Refrigerated and Insulated Vehicle

Highway transportation is the most important transportation mode in China's cold chain transportation, accounting for more than 90 % of total volume of cold chain transportation (Cold Chain Logistics Committee of China Federation of Logistics and Purchase 2015). In recent years, both holding volume and sales volume of refrigerated and insulated vehicles for highway transportation have grown substantially. In 2013, annual sales volume reached 14.0 thousand vehicles, with year-on-year growth of 63.8 %; annual holding volume reached 55.2 thousand vehicles, with year-on-year growth of 35.1 %, which is 30.3 percentage points higher than the growth of the holding volume of freight vehicles for highway transportation in the same period. Due to rapid growth in the holding volume, the proportion of refrigerated and insulated vehicles for highway transportation to freight vehicles for highway transportation was raised from 0.24 % in 2010 to 0.42 % in 2013. Total holding volume and growth rate of refrigerated and insulated vehicles in 2011–2013 are as shown in Fig. 12.5.

Table 12.4 Refrigerated warehouses built in 2013–2014 with storage capacity of 100 thousand tons or higher

Region	Refrigerated warehouse project- City/Name of enterprise	Storage capacity (thousand ton)
Liaoning	Dalian Port	100
Jiangsu	Nanjing/Yurun	100
Shandong	Weifang/Zhongkai	200
	Zhucheng/Longhai	400
	Qingdao	100
	Yantai/Ande	150
	Jining	100
Hubei	Yichang/Three Gorges	100
	Xiangyang/Sijiqing	100
Xinjiang	Urumchi/Haihong Distribution Center	100
	Urumchi/Haihong Logistics Port	100
Sichuan	Chengdu/Jingpin	100
	Chengdu/Yinli	100
	Chengdu/Haipawang	200
Chongqing	Chongqing/Cold Chain Logistics Center of Agricultural Products in Western China	200
Hebei	Langfang/Taigu	370 thousand cubic meters
Inner Mongolia	Baotou/Runheng	100
Jiangxi	Jiujiang/Jiangxi Haobo	200
Henan	Anyang/Topin	100
Yunnan	Kunming/Yunnan ASEAN Cold Chain Logistics Center	140

Source: Compiled *from the Report on China's Cold Cain Logistics Development of 2014, by Cold* Chain Logistics Committee of China Federation of Logistics and Purchase, China Logistics Technology Association, National Engineering Research Center for Agricultural Products Logistics, published by China Fortune Press, Beijing

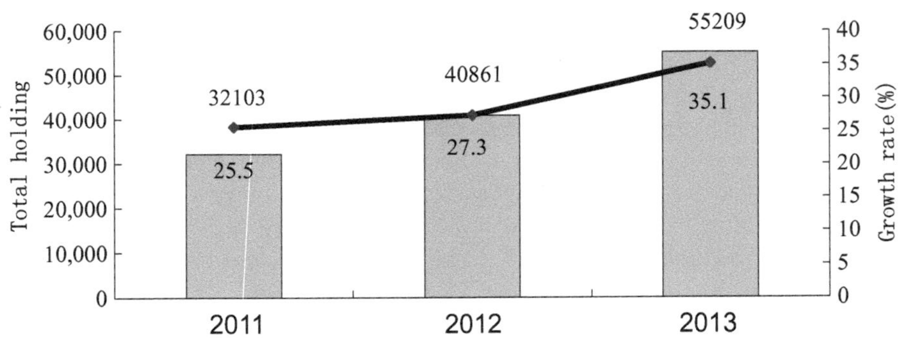

Fig. 12.5 Total holding volume and growth rate of refrigerated and insulated vehicles of China for 2011–2013
Source: Compiled from the Report on China's Cold Cain Logistics Development of 2014, by Cold Chain Logistics Committee of China Federation of Logistics and Purchase, China Logistics Technology Association, National Engineering Research Center for Agricultural Products Logistics, published by China Fortune Press, Beijing

12.3.3 Accelerated Development of Rail and Air Cold Chain Transportation

12.3.3.1 Rail Cold Chain Transportation

Items of rail cold chain transportation mainly include fresh perishable products. In recent years, China has accelerated its development of rail cold chain transportation. In 2009, China Railway Tielong Container Logistics Co., Ltd. successfully developed cold chain container for rail transport and suitable for multimodal transport. The company is mainly engaged in refrigerated container transport of cryogenic goods, such as quick-frozen foods, ice cream, and chilled meat, which need to be stored under ultralow temperature of −18 °C or below. Presently, it maintains 200 refrigerated containers, and has opened nearly 50 transportation lines and nearly 400 refrigerated container outlets, covering essentially all major provinces and cities in China.

In addition, in October 2013, China Railway Special Cargo Services Co., Ltd. and Nanning Railway Bureau, Guangxi Zhuang Autonomous Region instituted refrigerated cold chain transportation of fresh fruits and vegetables from Guangxi Bose to Beijing Dahongmen. After 107 hours on the road, 20 compartments of green vegetables, fruits and other goods from Bose were transported to Beijing; on the return trip, large amount of fruits and vegetables such as garlic and apples were brought back. The trip was the first time fruits and vegetables were exchanged between Guangxi and North China.

12.3.3.2 Air Cold Chain Transportation

Air cold chain transportation features fast speed, small volume, high freight cost and the need for multimodal transportation with other transportation modes. Items of air cold chain transportation in China mainly include low-temperature medicine, such as vaccines, blood products, and diagnostic reagents, which are generally of small size, high added value, and urgent time requirement, making cold chain transportation by air a suitable mode.

Recently, large cargo transportation airports and aviation companies in China have begun to regard cold chain logistics as a new growth spot in its cargo transportation development; items of air cold chain transportation have been expanded beyond medicine to fruits, flowers and aquatic products with high added value. For example, freight volume of flowers and fungi at Kunming Airport accounts for more than 80 % of its total cargo freight volume; flowers transported from Kunming to large and medium-sized cities in China would top 400 tons each day during holidays. For another example, as the southeast coastal region of China abounds in aquatic products, so regional airports there mainly handle shipments of fresh fish, shrimp and other sea products. In particular, at airports such as Weihai, Zhanjiang and Beihai, seafood products account for a relatively large proportion in airport freight volume.

12.3.4 Constantly Emerging Third-Party Cold Chain Logistics Enterprises

Third-party cold chain logistics enterprise is an important operating subject of cold chain logistics. Following the constantly increasing demand scale, a large group of professional cold chain logistics enterprises has emerged in China and played a significant role in promoting the scale and intensive development of cold chain logistics.

The first category is the transportation-type cold chain logistics enterprises which mainly engage in cold chain transportation and the business of trunk transportation and regional distribution. For example, Rokin Logistics Inc. owns over 450 vehicles for cold chain transportation and is engaged in carload and less-than-carload cold chain transportation on trunk and feeder roads. Over years of development, Rokin Logistics Inc. has formed a service network of cold chain logistics radiating throughout more than 20 provinces and cities in China, with first-tier cities such as Beijing, Shanghai, Guangzhou and Wuhan as the flagship. As another example, Express Channel Food Logistics Co., Ltd. is a third-party cold chain logistics enterprise mainly engaged in cold chain urban distribution and has a distribution network covering 43 domestic cities including Beijing, Kunming, Shenyang and Nanjing, serving clients including supermarkets, chain catering and fresh products e-commerce enterprises.

The second category is the storage-type of cold chain logistics enterprises which mainly engage in cold chain storage and the business of rendering low-temperature cargo storage, safekeeping, transfer and other services to clients. For example, Sino TransPFS, a cold chain storage logistics enterprise co-funded by Sinotrans Ltd. and the world fifth largest refrigerated warehouse operator Preferred Freezer Services, has constructed and operated three refrigerated warehouses at Lingang and Waigaoqiao in Shanghai and Tianjin Port, respectively. Among them, 40,000 ton refrigerated warehouse at Lingang has the largest auto racking system in Asia and can provide freezing and refrigeration services within the temperature range of −25 °C–10 °C. As another example, Jiangsu Runheng, a professional cold chain logistics enterprise mainly engaged in intelligent refrigeration of agricultural products and foods, has completed the construction of cold chain logistics centers and large supporting refrigerated warehouses in 15 provinces/cities. Its target is to build over 50 large modern food logistics parks covering the whole country by 2020.

The third category is the comprehensive-type cold chain logistics enterprises which mainly engage in integrated business such as cold chain storage, trunk transportation and urban distribution. For example, China Merchants Americold, a large public cold chain provider co-funded by China Merchants Logistics Holding Company Ltd. (CML) and Americold Realty Trust (Americold), has 14 refrigerated warehouses in major cities (with a total area of 150,000 m^2), over 100 self-owned refrigerator trucks, and integratively manages over 500 outsourced refrigerator trucks. Its transportation network covers major provinces and cities and can render integrated cold chain services of storage, transportation, distribution and other functions to clients. For another example, Shanghai Guangde, with freezing and refrigerated warehouse distribution centers in Beijing, Shanghai, Guangzhou, Chengdu, and Luoyang, occupying a total area of 80,000 m^2, can simultaneously provide long-distance and short-distance transportation services to first-tier and second-tier cities in China.

In addition, there is another category of cold chain logistics enterprises, which only offer urban distribution services at certain regions, with a relatively small service scope, though providing integrated services of cold chain storage and distribution. Examples of this type are Sinotrans Shanghai Cold-chain Logistics, and Shenzhen Shuguang Logistics.

12.3.5 Trend of Self-Constructed Cold Chain Logistics by Leading Enterprises

With the development of circulation system of fresh agricultural products in China, overall strength of manufacturing, wholesale and retail enterprises of agricultural products and food has been steadily bolstered; more and more attention has been directed toward the construction of auxiliary facilities for cold chain. Therefore, there has been an increasingly apparent tendency for large leading enterprises to independently construct their own cold chain infrastructures and run the cold chain distribution system. The advantage of self-built logistics is the relative flexibility in cold chain operations, which can be adjusted timely according to the enterprise's need in order to effectively ensure that storage, distribution and other operations are closely connected with internal demand, and thus improve the quality and efficiency of the cold chain services.

The first type of self-built cold chain logistics is by the processing and manufacturing enterprises. For example, Henan Topin Group, strategically positioned as "an integrator of Chinese meat industry chain," focuses on building a network service system of cold chain logistics that matches modern food processing and manufacturing industry. So far, a production and marketing logistics network has been constructed and put into operation, consisting of 10 cold chain logistics centers at the production origins and 15 cold chain logistics centers at the marketing sites. Based in Central China, the logistics network radiates throughout regions in East China, North China, Central-South China, and Northeast China.

The second type of self-built cold chain logistics is by the circulation and wholesale enterprises. For example, Fujian Mandy Group, one of the major wholesale market management enterprises of aquatic products in China, has been increasing its investment in constructing corresponding auxiliary cold chain facilities for the past few years. This group has successively invested in four large aquatic product trading centers in Fujian Province, Tianjin and Shandong Province, and promoted the operation system of modern cold chain wholesale market with a structural layout of "shop in front and warehouse behind." Among them, the construction project in Fujian Province is equipped with two sets of 50-kiloton refrigerated warehouses and five cold chain processing workshops; the construction project in Shandong Province is planned to have a 500-kiloton cold chain logistics park, in order to meet the demand of fresh products for cold chain storage, transportation and wholesale trade.

The third type of self-built cold chain logistics is by the chain retail enterprises. For example, the self-built fresh food processing and distribution center of Shanghai Lianhua Supermarket, covering a construction area of 35 km^2, can undertake the cold chain logistics distribution tasks of different fruits, vegetables and frozen foods. For another example, the fresh food distribution center built by Fujian Yonghui Supermarket in Fuzhou, with nearly 4,000 m^2 of refrigerated warehouse and over 20 self-operated refrigerator trucks, is capable of delivering 800 tons of fresh food daily to more than 100 chain stores in the province, within a distribution service radius of 300 km. According to statistics, among China's Top100 chain retail enterprises of fast moving consumer goods in 2013, 80 enterprises had built independent cold chain distribution centers of fresh foods, and most of them can distribute food to their stores two or more times each day.

Recently, some leading enterprises have actively expanded the social logistics service function in their cold chain logistics operations and gradually migrated toward third-party logistics operations mode. For example, in 2013, the Topin Group established the Xianyi Temperature-Controlled Supply Chain Co., Ltd. Founded on its scale and technical advantage in cold chain logistics, Topin provides socialized cold chain logistics and temperature-controlled supply chain services, including domestic and foreign trade, circulation and processing, temperature-controlled storage, trunk-line transportation, urban distribution, chain terminals, network marketing, and exhibition/transaction. Similarly, Shuanghui, Yurun, Brightdairy and several others have each established cold chain logistics company with an independent accounting system, to carry out the transformation from enterprise logistics to logistics enterprise.

12.3.6 Cold Chain Logistics of Fresh Product E-commerce Enterprises Becoming a Development Hotspot

Fresh product e-commerce enterprises face consumers through website platforms at marketing end and most of them make procurement at self-built base or wholesale market or via direct purchase from the producing origins at supply end. Therefore, cold chain logistics has become a key link to connect supply and marketing ends. Especially, cold chain home delivery, as "the last mile" of fresh product e-commerce logistics, will directly serve consumers; its level of development has become the bottleneck factor which determines whether the fresh product e-commerce enterprise can survive or grow.

At present, fresh product e-commerce enterprises in China include: (1) integrated platform e-commerce enterprise with fairly complete product categories, such as Tmall.com, Jd.com, Yhd.com, Suning.com, and Amazon.cn; (2) vertical e-commerce enterprise specialized in food e-retailing, such as Womai.com, Tootoo.cn, Benlai.com, Yoocai.com, and Yiguo.com; (3) fresh product e-commerce enterprise managed by logistics enterprise, such as Sfbest.com. These different fresh product e-commerce enterprises often choose different cold chain logistics modes according to their own features, as shown in Table 12.5.

Platform-type of e-commerce enterprises often adopts third-party cold chain logistics. For example, Tmall.com expands its fresh product market by integrating its offline cold chain logistics resources through Cainiao Logistics Platform. Specifically, Tmall chooses excellent cold chain logistics enterprises in the cities within its key business scope as the designated cold chain logistics operators of Tmall.com, and integrate them together to serve fresh product suppliers on the Cainiao Logistics Platform. Some channels adopt the one-stop whole-process cold chain distribution while others adopt the "two-stage cold chain distribution" mode. In the latter mode, intercity trunk transportation is carried out through B2B refrigerator trucks, and then the B2C home delivery in "the last mile" is achieved at normal temperature or with a refrigeration box. Similarly, Suning.com and Jd.com make cold chain operation for their open platform-based fresh product channel through logistics cooperation with the third party. "Yhd.com," in addition to marketing of fresh products, also builds special fresh product channel to independently market its fresh products; it makes whole-process cold chain distribution of fresh products through supplier review, product check-in inspection, storage and distribution management, settlement and tracing of after-sales quality problems and other supervision and safeguarding measures. Furthermore, Yhd.com also adheres to the prompt delivery standard of guaranteeing merchandise arrival within 24 hours after dispatching from the warehouse; however, its business scope is only limited to Beijing and Shanghai at present time.

Vertical e-commerce enterprises are predominately engaged in fresh products and often adopt the self-built cold chain system or a mixed mode of "self-built system + third-party system." For example, Tootoo.cn built its own logistics fleet in the beginning of its establishment and now adopts the mode of "refrigerator truck + tricycle + refrigeration box" with strong regional characteristics to render cold chain distribution service to Beijing consumers. It adopts the mode of logistics cooperation with third party in fresh product business outside Beijing, effectively controlling cold chain cost and improving distribution flexibility. For another example, "Benlai.com" adopts self-built storage mode and also cooperates with the third-party logistics enterprise, vtepai, in order delivery.

Sfbest.com, relying on the third-party logistics enterprise SF Express, has unique advantages in logistics. Now it has built three cold chain warehouses, namely the North China warehouse, East China warehouse and South China warehouse. Each warehouse consists of five sections for maintaining the freezing (−18 °C), refrigeration (2–8 °C), constant temperature (15 °C–25 °C) and other temperature-controlled storage functions. It fulfills the transportation function by relying on SF Express as

Table 12.5 Product category, cold chain logistics mode and distribution scope of some fresh product e-commerce enterprises

E-commerce enterprise classification	Enterprise name	Business mode and product category	Cold chain logistics mode	Distribution scope of cold chain
Platform	Tmall.com	Open platform, dealing in aquatic products, meat, fresh fruits and vegetables; special fresh product channel, for advanced sale of distinctive imported products.	Cooperation with third-party logistics enterprises; whole-process cold chain distribution in some channels; other channels adopt the "B2B + B2C" two-stage cold chain distribution mode, which means supplementing third-party logistics enterprise in cold chain B2C home delivery, based on B2B cold chain intercity transportation	42 domestic cities
	Jd.com	Open platform, dealing in eight product varieties, including seafood products, fruits, vegetables, eggs, fresh meat, etc.	Cooperation with third-party logistics enterprises: S.F. Express and Express Channel Food Logistics, in the mode of whole-process cold chain or "B2B + B2C" two-stage cold chain home delivery	Beijing
	Suning.com	Open platform, dealing in six product categories: including seafood products, quick-frozen foods, processed meat, eggs, etc.	Cooperation with third-party logistics enterprise; with possibility in self-built cold chain warehouse and logistics distribution system	58 regions in China
	Yhd.com	Self-run No.1 fresh food channel; open platform, dealing in three product categories: fruits, vegetables and chilled foods.	Logistics of self-run business undertaken by Wal-Mart distribution system or third-party logistics; open platform managed by suppliers; expected to independently build cold chain distribution channels in the future when the number of daily order reaches 10,000	Beijing, Shanghai

Integrated e-commerce enterprise	Womai.com	Dealing in fresh products of 18 categories, including fruits, live poultries, eggs, etc.	Both self-built logistics system and third-party logistics. Self-built refrigerated warehouse in North China and other regions and cooperation with Huazhengdao Logistics, Beequick, and other third-party transportation and distribution enterprises	Seven provinces and cities, including Beijing, Shanghai, Guangzhou, Tianjin, etc.
	Tootoo.cn	Dealing in 16 product categories of fresh products, including organic fruits and vegetables, seafood, snacks and drinks, fresh meat, etc.	Whole-process cold chain distribution with self-run logistics channel in Beijing and whole-process cold chain with third-party logistics channel in other regions	32 domestic cities
	Benlai.com	Dealing in fresh products of four categories, including fruits and vegetables, meat, poultry and eggs, dairy products, and imported food, etc.	Self-built warehouse or cooperation with third-party logistics enterprise vtepai to realize whole-process cold chain distribution	Seven provinces and cities including Beijing, Tianjin, Guangdong and Shanghai, etc.
Managed by express company	Sfbest.com	Dealing in fresh products of eight categories, including meat, seafood and aquatic products, diary drinks, fruits, vegetables, cooked meat food, staple pasta, instant ice product, etc.	Self-built cooling and refrigerated warehouse in Beijing, Guangzhou, Jiaxing and other locations; carry out one-stop whole-process cold chain distribution with the service of S.F. Express	Expanded to 11 cities from Beijing and Tianjin

Source: Compiled from data disclosed by major e-commerce enterprises

a strong logistics system support, to implement the whole-process cold chain distribution of "production origin to dining table," and uphold its positioning in high-end fresh product marketing. For example, Sfbest.com sells litchi gift box by adopting the mode of direct procurement from the production origin, whole-process temperature-controlled distribution, arrival within 24–30 hours, and traceability system with QR code tag to ensure product quality and favorable consumer consumption experience. Hence, the litchi product still generates high sales proceeds, though its price is far higher than that in regular markets.

In addition, through innovative modes such as building of fresh product self-service box, designated supermarket collecting agency and self-built service station, e-commerce cold chain home delivery has been carried out more efficiently and the problem of inconvenience in receiving delivery has been solved. For example, fresh product e-commerce enterprise justeasy.com.cn sets up temperature-keeping cabinets in residential communities for electronic fresh product ordering boxes. It sends delivery message to consumers after distributing and delivering the products as ordered by consumers, who can then go to the designated electronic ordering box to pick up their products using membership card or SMS password. Nowadays justeasy.com.cn will make product distribution to the communities with electronic ordering box twice a day, and each distribution courier may deliver more than 300 orders on each trip. For another example, Jd.com cooperates with Taiyuan Tangjiu convenience store to provide fresh product pick-up stations to community residents covered by over 600 outlets of Tangjiu convenience store in the city. For another example, Tmall.com has established Tmall.com service stations in 25 cities across the country, including Beijing, Shanghai, Guangzhou, Shenzhen and Nanjing, which are used to provide pick-up service to customers in offline physical stores.

12.4 Existing Problems and Development Trend of Cold Chain Logistics in China

In recent years, though China has made notable progress in cold chain logistics development, it still has some problems needing prompt solution in infrastructure construction, level of logistics enterprise, and technology promotion and application. Along with development of cold chain industry and improvement of supporting environment, construction of cold chain logistics infrastructures shall be continually strengthened, cold chain logistics enterprises shall be further developed and cold chain logistics technology shall be improved.

12.4.1 Existing Problems in Cold Chain Logistics

12.4.1.1 Insufficient Cold Chain Infrastructure Capacity

Though China has made great progress in construction of cold chain logistics infrastructures, yet it is still insufficient to meet the ever-increasing need of cold chain logistics. So far, China's per capita refrigerated warehouse capacity is about 0.1 m^3, up to the medium market level of refrigerated warehouse in emerging market countries,[3] but far lower than that of developed countries (for example, 0.36 m^3 in US and 0.33 m^3 in Japan). The number of refrigerator trucks accounts for about 0.42 % of total number of freight vehicles, far less than that of developed countries (for example, 0.8–1 % in US, and 2–3 % in Germany) (Cold Chain Logistics Committee of China Federation of Logistics and Purchase 2015). In addition, there exists marked difference among regions in the construction of cold chain logistics infrastructures; the Central and Western regions and rural regions in China lag severely behind in construction of cold chain logistics node, constricting the improvement of integrated service network of cold chain logistics.

12.4.1.2 Strength of Cold Chain Logistics Enterprises Needing Improvement

Cold chain logistics enterprises develop rapidly in China, but their overall level still needs further improvement. First, cold chain enterprises have relatively small business scale and relatively low degree of market concentration. In 2013, gross operating income of Top 100 Cold Chain Enterprises was 10.90 billion RMB, which accounts for merely 10 % of the gross income of all national cold chain enterprises; overall there were only ten or more enterprises which had an operating income over 100 million RMB (Cold Chain Logistics Committee of China Federation of Logistics and Purchase 2014b). Secondly, cold chain service coverage is relatively small and professional service level needs improvement. Most enterprises confine their cold chain distribution to a relatively small area, mainly offering only traditional transportation and storage service. Most of them lack the service capability in high value-added businesses, such as cold chain circulation and processing, whole-process cold chain control, and temperature-controlled supply chain solutions. These deficiencies result in serious homogeneous competition and increasingly compressed profit margins among them. Thirdly, some enterprises are disadvantaged due to low cold chain technology, low management ability and low service level. For example, it is difficult for some cold chain logistics enterprises to meet the requirement of cryogenic products in temperature control, and there still exist

[3]According to the Global Cold Storage Capacity Report 2012 of the International Association of Refrigerated Warehouses, China has reached the medium market level of refrigerated warehouse among emerging countries.

the phenomena of "mixed loading of cold and hot products" and "cold chain breakage" in distribution links.

12.4.1.3 Lagging Application of Cold Chain Logistics Technology

China's cold chain logistics technology needs further enhancement. Especially, the promotion and application of certain advanced cold chain technologies are still at the start-up stage; commercial processing means, such as post-production pre-cooling technology and classification and packaging and processing under low temperature environment, have not been popularized in the industry; temperature control measures in transportation are primitive and crude; whole-process automatic temperature control technology widely applied in developed countries is still only applied in some parks and enterprises for pilot application, not widely industrialized. Moreover, modern information technology is not fully implemented. As a result, on the one hand, enterprises' IT application in cold chain logistics management is low and it is difficult to realize real-time monitoring and quality tracing of cold chain logistics products; on the other hand, the construction of public information platform of cold chain logistics is lagging, so it is difficult to optimize allocation of logistics resources through integration of cargo and cold chain vehicle information.

12.4.2 Development Trend of Cold Chain Logistics

12.4.2.1 Construction of Cold Chain Infrastructure to Be Further Strengthened

Demand for fresh agricultural products and food in China will continue to grow and Chinese governments will continue to increase their financial support for cold chain industry. As influenced by the demand and policy, China will further expand its construction scale of cold chain logistics infrastructure and improve its service capability in the future. The construction of large public refrigerated warehouses will continue to grow on the whole; small and medium-sized refrigerated warehouses at production and marketing end will unleash the development potential. Leading enterprises in circulation and processing, logistics, e-commerce and other fields will accelerate their actions in cold chain logistics. At the same time, more efforts will be made to construct temperature-controlled facilities in processing links of agricultural products and economical and suitable pre-cooling facilities of agricultural products; the inspection and testing facilities and devices complying with cold chain logistics will be promoted and applied.

12.4.2.2 Strength of Professional Cold Chain Enterprises Needing Improvement

Cold chain logistics enterprises have started to expand their scale and improve their market share through equity sharing and holding, merging and reorganization, cooperation and alliance. For example, in 2013, Shanghai Zhengming Logistics announced that it had obtained private fund of 120 million RMB for investment and taken over Shenzhen Shuguang Logistics. With increasingly enlarged enterprise scale and professional capability, service of cold chain logistics enterprises will not only be limited to traditional cold chain storage and distribution, but also be expanded to deeper cooperation with clients in cold chain circulation and processing, monitoring and tracing, temperature-controlled supply chain solutions and other value-added services. High value-added, individual and integrated cold chain logistics services will certainly become a business development direction of cold chain logistics enterprises in the future.

12.4.2.3 Cold Chain Logistics Technology Needing Further Improvement

Independent research and development, introduction, digestion and absorption of new cold chain logistics technologies will be further promoted, mainly including the research and development of high-performance refrigeration, sorting, cleaning, processing and packaging facilities and technology, economic and applicable pre-cooling facilities of agricultural products, mobile cooling devices, energy-efficient cold chain transportation tools, advanced display and sales facilities and technologies, cold chain monitoring and tracking system, and temperature-controlled facilities. In addition, the *Guidance on Further Promoting Healthy Development of Cold Chain Transportation Logistics Enterprises* issued by the NDRC in 2014 pointed out, "to actively promote information construction of cold chain transportation logistics and strengthen the application of the Internet of things, cloud computing, big data, mobile Internet and advanced information technologies in cold chain transportation logistics." Therefore, to promote innovative development of cold chain logistics on the basis of high-new information technologies is also one of the development directions of cold chain logistics in the future.

12.5 Summary

This chapter discusses the development environment, development status and existing problems of cold chain logistics in China in 2014 and its development trend for the future. In terms of development environment, Chinese residents' consumption level has been continually elevated, fresh product e-commerce has been rapidly developed and production scale of cold chain products has been continuously expanded. As a result, relevant policy environment has also been improved with time. In terms of development status, the demand of cold chain logistics has been notably increased, the infrastructure construction has been greatly enhanced, the development of cold chain logistics by rail and air has been accelerated and the third-party cold chain logistics enterprises, leading production and circulation enterprises and cross-border e-commerce enterprises have paid more attention to cold chain logistics construction. Recognizing that it still has some problems in cold chain logistics development, China will continue to strengthen the construction of cold chain infrastructure, and its professional cold chain logistics enterprises will further enhance their strength and improve their cold chain logistics technologies as time goes on.

References

Alibaba Group Research Center (2014) The Report of the Alibaba Agri-product E-commerce of 2013, published by Alibaba Group Research Center. http://www.aliresearch.com/blog/article/detail/id/18882.html

Cold Chain Logistics Committee of China Federation of Logistics and Purchase (2014a) Report on China's Cold Chain Logistics Development of 2014, Cold Chain Logistics Committee of China Federation of Logistics and Purchase, China Logistics Technology Association, National Engineering Research Center for Agricultural Products Logistics 2014, China Fortune Press, Beijing

Cold Chain Logistics Committee of China Federation of Logistics and Purchase (2014b) The top hundred China cold chain logistics corporations of 2013, Cold Chain Logistics Committee of China Federation of Logistics and Purchase, Beijing. http://www.lenglian.org.cn/xhdt/20610.shtml. Accessed 17 Dec 2014

Cold Chain Logistics Committee of China Federation of Logistics and Purchase (2015) General Review of China's Cold Chain Logistics Development of 2014. Cold Chain Logistics Committee of China Federation of Logistics and Purchase. Beijing. http://www.snet.com.cn/109/2015_3_24/3_109_306124_1776_0_1427183405875.html. Accessed 25 Mar 2015

Haiyan Y (2007) China agri-product cold chain logistics analysis and system optimization. Transp Preserv Commod 6:49–50

National Bureau of Statistics of China (2014) The China Statistical Yearbook (2013), China Statistical Bulletin of National Economic and Social Development (2013), published by the National Bureau of Statistics of China, Beijing, 24 Feb 2014

Kolibri 360 GmbH
Ferdinand-Jühlke-Straße 7
99095 Erfurt, Deutschland
produktsicherheit@kolibri360.de